Students' Motivations and Emotio[Chinese Science Classrooms

The book reviews and examines students' motivations and emotions in Chinese science classrooms.

By adopting different approaches such as content analysis, factor analysis, path analysis, and latent profile analysis, the author analyzes the content of literature, curriculum standards and textbooks, classroom observations, survey data, interview data, and open-ended responses from students and teachers through a literature review and six empirical studies. The findings may provide insights for education researchers and practitioners seeking to improve science teachers' pedagogical practices and create friendlier classroom environments.

Researchers of science education or those who are interested in investigating students' affective perceptions in specific subject contexts will find this book interesting.

Xiaoyang Gong is an assistant professor at Faculty of Education, Tianjin Normal University. Her research interests include science education, psychology, and education technology.

Students' Motivations and Emotions in Chinese Science Classrooms

Xiaoyang Gong

Routledge
Taylor & Francis Group

LONDON AND NEW YORK

This material is based upon work supported by the Youth Project of the Ministry of Education Research Fund under grant number EHA230487.

First published 2024
by Routledge
4 Park Square, Milton Park, Abingdon, Oxon OX14 4RN

and by Routledge
605 Third Avenue, New York, NY 10158

Routledge is an imprint of the Taylor & Francis Group, an informa business

British Library Cataloguing in Publication Data
A catalogue record for this book is available from the British Library

ISBN: 978-1-032-64961-0 (hbk)
ISBN: 978-1-032-64970-2 (pbk)
ISBN: 978-1-032-64966-5 (ebk)

DOI: 10.4324/9781032649665

Typeset in Times New Roman
by SPi Technologies India Pvt Ltd (Straive)

Contents

Figures

Tables

Part 1

Introduction

1 Introduction to Motivations and Emotions in Chinese Science Classrooms

Introduction

It is widely acknowledged that students' motivations and emotions play important roles in academic settings (Eccles, 2005; Pekrun, 2006). They are closely associated with students' cognitive antecedents and outcomes (Ashby, Isen, & Turken, 1999). As students' perceptions of motivations and emotions are domain specific (e.g., Guo, Marsh, Parker, Morin, & Dicke, 2017; Nagy, Trautwein, Baumert, Köller, & Garrett, 2006) and shaped by contextual factors (e.g., Kwon, Yoon, Joormann, & Kwon, 2013), there has been growing interest in investigating these variables in different domains and across different cultural contexts. Yet, the mechanisms of how Eastern students' motivations and emotions influence their academic outcomes in the science domain remains under-researched (Flaherty, 2020), which calls for more research to capture how motivations and emotions function in this specific context. To address this issue, the first chapter adopts a qualitative content analysis approach to examine how motivational and affective dimensions are represented in the Chinese K-12 science curriculum and summarize the main themes that emerged from the research literature of the past decade published in Chinese. I hope that this effort may inspire more science education researchers and educational psychologists to explore this interdisciplinary research area more deeply and creatively. This chapter includes three parts: the first part describes the disciplinary and cultural context of this study; the second part analyzes the representation of motivations and emotions in Chinese K-12 science curriculum; and the third part reviews related literature and summarizes main findings. Finally, a series of implications for research and practice are suggested.

Science Education in Eastern Contexts

Before conducting the qualitative content analysis, it is important to understand the context of this study – science teaching and learning in Eastern contexts. Science was imported into China at the beginning of the 19th century (Abd-El-Khalick et al., 2004). As science has evolved from Western cultures, teaching and learning science in Eastern contexts is a complex undertaking

DOI: 10.4324/9781032649665-2

that involves the convergence of different cultural values. For example, the Western science knowledge tells students that the Earth is round while the ancient Chinese yin-yang theory implies a basic notion of round sky and square earth (*Tian Yuan Di Fang*). Such contradictory viewpoints derived from different cultures increase science teachers' difficulties in explaining scientific concepts or helping students achieve conceptual change. Figure 1.1 illustrates the conflicts between traditional cultures and scientific views of man and nature and ways of thinking (Ogawa, 1986). Just as Baker and Taylor (1995) described, the simplified attempt to replicate and implement Western science curriculums appears to be ineffective due to its mismatch with students' world views, cultural beliefs, and learning needs in non-Western countries. As a consequence, the format and process of science education in Eastern contexts are quite different from those in Western contexts due to the influence of cultural values or norms defined in such environments.

Two dominating cultures – Confucianism and Collectivism – delineate the public's perceptions about responsibilities of teachers and students in Chinese school settings. As Confucian cultures emphasize education's function of helping students achieve a higher socioeconomic status, developing moral characteristics, and preserving the family's honor, a tradition of deep respect for teachers has been established among Chinese student populations (Ma, 1999). Teachers, who are usually called "engineers of the soul", are authority figures second to parents (Wang, Wang, Zhang, Lang, & Mayer, 1996). The old saying "once my teacher, forever my parents" is commonly used to educate students to respect teachers. In turn, teachers should fully support students in acquiring content knowledge and getting high scores in school examinations. Due to the emphasis on knowledge transmission, student academic performance is regarded as the most important criterion for evaluating a teacher's effectiveness and a school's quality. Under such cultural and value norms, the Chinese education system is highly centralized with national curriculum standards, textbooks, and examinations. For example, the high school entrance examination in ninth grade (i.e., the last year in middle school) and the college entrance examination in 12th grade (i.e., the last year in high school) are regarded as the two remarkable events in one individual's academic career, which directly

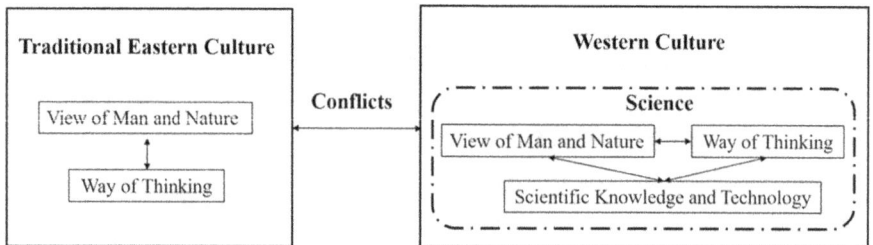

Figure 1.1 Ogawa's proposed model for a rationale of science education.
(Ogawa, 1986).

determine his or her future development. Under the big pressure of public examinations, the traditional science instruction is teacher-centered, theory-focused, examination-oriented (Wang et al., 1996) and supplemented with demonstration experiments, videos, and homework. The underlying reason for this lecture-dominated instruction is that it maximizes the efficiency of teaching and learning because students receive rich knowledge and information within the limited class time. Teachers usually interact with students in the format of choral responding. Consequently, there are few opportunities for students to interact with each other during class periods. Students heavily rely on teachers' explanations to understand scientific knowledge and seldom interrupt teachers' lectures to ask questions, which may take away time for the whole class. As classroom activities are teacher directed, student engagement in Chinese classrooms refers to the continuous retainment of attention toward teachers' directions and explanations. It should be noted that the above situation has considerably improved in recent years, specifically in developed regions such as Beijing and Shanghai due to national-level educational reforms through publishing or revising K-12 curriculum standards and school textbooks, which serve as the central pillar for science instruction. In contrast, Western classroom learning environments are more participatory and active where students can ask the teacher questions directly in whole-class instruction. Students are also encouraged to interact with classmates in the format of classroom discussions and small-group activities. The differences in classroom environments (independence versus collaboration) and cultural values (Collectivism and Confucianism versus Individualism) potentially influence students' motivations for school learning and experiences and expressions of emotions.

Just as described previously, the Ministry of Education revised high school curriculum standards in 2020 and published new compulsory education curriculum standards in 2022 (The Ministry of Education, 2020a, 2020b, 2020c, 2022a, 2022b, 2022c, 2022d) due to the increasing criticism that the over-emphasis on content knowledge may lead to the passive rote memorization and decreased students' interests in learning science. These national endeavors suggest reforming the traditional lecture-based instruction and advocating student-centered holistic education. School curriculum standards in each domain delineate specific expectations for cultivating students' core literacies, which are defined as correct values, necessary qualities, and key competences formed in the process of learning one specific subject. The ultimate goal is to promote humanistic development and scientific literacy of the whole population.

Motivations and Emotions in K-12 Science Curriculum Standards in China

In the Chinese K-12 education system, the science subject exists in different forms across different grade levels. As shown in Table 1.1, elementary school students learn the science subject from first grade to sixth grade. Starting from

Table 1.1 Science-related subjects offered at different levels

School Levels	Compulsory Education		High School
	Elementary School	Middle School	
Grade Levels	1st–6th grade	7th–9th grade	10th–12th grade
Subjects	Science	Physics, biology, and chemistry (or science)	Physics, biology, and chemistry

middle school, students can learn science as a whole or learn its alternative branches including physics, biology, and chemistry. The compulsory curriculum plan suggests that science-related courses should account for eight to ten percentage points of the nine-year compulsory-education class periods. In high school, the science subject no longer exists as a whole and students only learn three branch subjects of science (i.e., physics, chemistry, and biology). They learn the three subjects simultaneously in tenth grade. After one year of learning, 11th-grade students can select three out of six subjects (i.e., physics, chemistry, biology, politics, history, and geography) as subjects tested in the national college entrance examination, which may be related to their major or career choices in the future. The newly published compulsory education science standards and revised high school science standards provide rich sources for understanding how motivations and emotions are represented in science curriculums at the national level.

Compulsory Education Science Standards

As described in the compulsory education science standards, the goal of science education in the compulsory stage is to cultivate students' four types of core literacies including (1) *scientific ideas*; (2) *scientific thinking*; (3) *scientific inquiries and practices*, and (4) *scientific attitudes and social responsibilities*. Among them, *scientific ideas and scientific thinking* describe the cognitive dimension of learning science and *scientific inquiries and practices* emphasize the practical dimension while *scientific attitudes and social responsibilities* are closely related to the motivational and affective dimension of learning science. *Scientific attitudes*, which encourage students to be realistic, practical, and serious in scientific research, are represented in diverse scenarios. For example, students keep continuous curiosity and enthusiasm for exploring scientific questions, think rigorously and logically, express their opinions with evidence, enjoy the process of collaboration and communication, pursue creativity, and oppose blind faith in high authority. *Social responsibilities* are embodied in students' behaviors and actions in daily lives. For example, students treasure life and live a scientific and healthy lifestyle; students love nature, save resources, protect environments, and promote sustainable development and ecological civilization; students make right decisions in the face of hot social scientific or technological issues; students obey legal, social, and moral rules during the application of science and technology; and students protect rights of the self and others and safeguard national interests.

Besides compulsory education science standards, the core literacy of *scientific attitudes and social responsibilities* is also represented in three branches of science offered in middle school. In the physics domain, students can understand the nature of science, seek truth from facts, and show great courage and perseverance. In the chemistry domain, students have deep interests in learning chemistry and scientific inquiry; appreciate the value of chemistry in facilitating human civilization and sustainable development; and love the country and shoulder the responsibility of reviving the Chinese nation and driving social development. In the biology domain, students form the disposition of open-mindedness and inclusiveness. These examples in national compulsory curriculum documents elaborate on the role of science-related subjects in influencing students' motivations and affections positively at both the individual level and the national level.

High School Curriculum Standards of Physics, Biology, and Chemistry

Similar to compulsory education science standards, high school curriculum standards of physics and biology share four types of core literacies including physical or biological ideas, scientific thinking, scientific inquiry, social responsibility, and scientific attitude (see Table 1.2). Such similarity results from the effort of revising compulsory education curriculum standards which aims to make it more coherent and consistent with high school science education. Compared to high school standards of physics and biology, chemistry curriculum standards list five core literacies including the exploration and analysis at macroscopic and microscopic levels, the notion of change and balance, using evidence to make inferences, modeling, scientific inquiry, creativity, scientific attitude, and social responsibility. The representation of motivations and affections in core literacies are shown in italic texts. Examples of representations are listed in the last column of Table 1.2, including attitudes (e.g., the nature of science), value beliefs, intrinsic motivations, and so on.

In summary, promoting students' self-efficacy beliefs, interests, and motivations for learning science is one important curriculum goal in K-12 science education, which describes the interaction between the student self and the science subject. Simultaneously, Chinese science curriculum standards also highlight the importance of developing the spirit of patriotism, the sense of national pride, and the mission of rejuvenating the nation, which indicates the interaction between the student self and the whole nation. At present, schoolteachers are recommended to integrate ideological and political education in different school disciplines. For example, science teachers can introduce the glories of ancient China (e.g., the four great ancient inventions) and tell stories of great scientists with modern scientific achievements (e.g., Hou Debang, the inventor of Hou's process for soda production). The aforementioned curriculum value

Table 1.2 Core literacies listed in high school standards of physics, biology, and chemistry

Subjects	Core Literacies	Motivations and Emotions
Physics	• Physical ideas • Scientific thinking • Scientific inquiry • *Scientific attitude and responsibility*	• The nature of science; • Intrinsic motivations for exploring nature; • Attitudes of persistence and realism.
Biology	• Ideas of life • Scientific thinking • Scientific inquiry • *Social responsibility*	• Ecological consciousness; • The involvement in social discussions; • The admiration of healthy and civilized lifestyles; • The contribution to building a healthy China.
Chemistry	• Exploration and analysis at macroscopic and microscopic levels • The notion of change and balance • Using evidence to make inferences and modeling • Scientific inquiry and *creativity* • *scientific attitude and social responsibility*	• The safety consciousness; • The aspiration to seek the truth; • The pursuit of a green and low-carbon lifestyle; • Value beliefs in chemistry.

orientation illustrates how Eastern cultural values (e.g., Collectivism) shape the development and implementation of science curricula in China.

Motivations and Emotions in Chinese Literature from 2010 to 2020

Besides compulsory education science standards and high school science standards, the research literature was also examined to identify main themes and therefore provide suggestions for future directions. In this section, I address the need for reviewing articles that have examined students' motivations and emotions in the science domain between 2010 and 2020.

Data Collection and Qualitative Content Analysis

The literature review was processed in three phases. In the first phase, I searched for relevant articles from three main databases in China: China National Knowledge Infrastructure (CNKI, https://www.cnki.net/), China Science and Technology Journal Database (http://lib.cqvip.com/), and Wanfang Database (http://www.wanfangdata.com.cn/) with several key words including attitudes, interests, achievement emotions, self-efficacy beliefs and motivations. In the second phase, I narrowed down the list of articles and only focused on articles discussing students' motivations and emotions in the science domain. Studies in other areas (e.g., math) or research on teacher populations (e.g., pre-service teachers and in-service teachers) were excluded. It should be noted that the paradigm of conducting education research differs across countries and over time. In China, the traditional research paradigm in different subject domains includes the following aspects: improving instructional practices (e.g., developing lesson plans and analyzing exam questions), deepening theoretical understanding (e.g., outlining advanced education theories), and enhancing the connection between theory and practice (e.g., providing implications). In recent years, the empirical research paradigm is recommended and the number of empirical studies has increased. Yet, education research in specific disciplines is considered as a branch of research in curriculum and instruction, which is also a branch of education research as a whole. This notion leads to the consequence that the number of related academic journals is relatively small. For example, there are only three main journals directly related to chemistry education: *Chinese Journal of Chemical Education* (化学教育), *Education in Chemistry* (化学教学), and *Teaching Reference of Middle School Chemistry* (中学化学教学参考); two journals are directly related to biology education, including *Biology Teaching* (生物学教学) and *Teaching of Middle School Biology* (中学生物教学); several journals related to physics education include *Teaching Reference of Middle School Physics* (中学物理教学参考), *Middle School Physics* (中学物理), *Physics Teachers* (物理教师), *Physics Teaching* (物理教学), and *Physics Experimentation* (物理实验). Taking the number of schoolteachers and teacher education researchers into consideration, there are limited spaces for them to publish research articles in academic journals. In this literature review, a total of 37 articles published between 2010 and 2020 were included in the final review (see the Appendix). They covered theoretical papers, literature reviews, and empirical studies. Among them, 17 articles were indexed in the Chinese Social Sciences Citation Index (CSSCI) and Chinese Core Journal Catalogue (CCJC), which are more likely to have higher citation rates than general journals. The purpose of this literature review is to answer the following research question: what themes and research approaches dominate studies of motivations and emotions in the science domain in China?

After identifying articles that met the inclusion criteria, in the third phase, I conducted the qualitative content analysis aiming to categorize the previously mentioned studies based on the content review procedure (Fraenkel, Wallen, & Hyun, 2015). Four main themes were applied: (1) disciplinary contexts, (2) variables being investigated, (3) research methods being used, and (4) research questions and findings. An Excel spreadsheet was used to list codes or detailed information extracted from each article that was related to the research question. Regarding the disciplinary contexts, the distribution of reviewed articles was as follows: 14 articles (37.84%) in chemistry, seven articles (18.92%) in biology, 13 articles (35.14%) in physics, and three articles (8.1%) in elementary science. Regarding paper formats or employed research methods, there were four literature reviews, 11 theoretical papers, and 22 empirical studies that analyzed survey data quantitatively. Among 22 empirical studies, five studies focused on self-efficacy, two studies on achievement emotions, six studies on motivation, and nine studies on multi-motivational or emotional constructs. In the following section, I will describe these motivational or emotional variables being investigated in sequence.

Science Self-Efficacy Beliefs

Science self-efficacy is defined as one individual's belief in his or her ability to organize and execute the courses of action required to successfully complete a science-related task (Bandura, 1986; Pajares & Miller, 1994). Students who perceive different levels of self-efficacy may exhibit different patterns when facing challenges and difficulties in science-related tasks: students with high self-efficacy are more likely to insist and spend more effort to figure out the solution while students with low self-efficacy may give up and doubt themselves. As one individual's self-efficacy belief plays an important role in predicting his or her academic performance, recent research has focused on developing various instruments and identifying its relationship with other variables, which provide implications for teachers' pedagogical practices.

As different theoretical frameworks define the construct of self-efficacy beliefs in different ways, authors of reviewed articles developed or used different instruments to measure students' science self-efficacy beliefs. For example, the action theory conceptualizes students' self-efficacy as beliefs in perceived control regarding relations between agents, means (or causes), and ends (or outcomes) of actions (Skinner & Chapman, 1984; Skinner, Chapman, & Baltes, 1988). Using the action theory as the theoretical framework, Dong (2015) adapted items from existing surveys (e.g., Bian, 2003; Schwarzer & Jerusalem, 1997) and measured 12th-grade students' chemistry self-efficacy with 48 items categorized in three dimensions: (1) perceived self-regulation of chemistry attitudes; (2) perceived self-regulation of chemistry learning behaviors; (3) perceived self-prediction of chemistry learning outcomes. Specifically, the first dimension included students' perceived talent, perceived self-doubt, perceived control of emotions, and perceived effort; the second dimension included students' perceived self-monitoring, perceived helplessness, and perceived control of attention; the third dimension included positive self-expectation, beliefs of getting good grades, self-assurance and perceived goal attainment. Similarly, Li, Fan, and Tao (2012) measured high school students' chemistry self-efficacy in four dimensions with 26 self-developed items. The first three dimensions were derived from the action theory and included students' perceived self-assurance, self-doubt, and goal attainment. Besides these terms, they added the fourth dimension – the efficacy of doing hands-on experiments.

However, the lack of detailed information about the procedure of developing survey items threatens the content and construct validity of these self-developed survey items. In contrast, Chen and Bi (2011) used the attribution theory (Weiner, 1986) as the theoretical framework and measured chemistry self-efficacy with 22 items categorized into two aspects: perceived learning abilities and perceived learning behaviors. Other researchers were more likely to adopt the social-cognitive theory (Bandura, 1986) as the theoretical framework. For example, Liu (2019) adapted eight items from the Motivated Strategies for Learning Questionnaire (MSLQ; Pintrich et al., 1991) and other surveys (Wang & Lin, 2007) to measure elementary students' self-efficacy in learning science. In addition, there was one study measuring high school students' physics self-efficacy with 35 self-developed items without describing the concrete procedure (Hu, Zhang, & Wu, 2010).

Besides the general measurement of self-efficacy beliefs in the science domain, some researchers tried to evaluate specific types of self-efficacy (e.g., inquiry self-efficacy and collective self-efficacy). For example, Zhang, Dong, Bai, Xiong, and Zhu (2018) adapted 15 items from McGill's Self-Efficacy of Learners for Inquiry Engagement (Ibrahim, Aulls, & Shore, 2016) and measured elementary students' self-efficacy of inquiry in three aspects: students' personality characteristics, carrying out practices of science and achieving inquiry-learning outcomes (e.g., understanding, application, evaluation, and creation of content knowledge). Besides the personal-level science self-efficacy mentioned previously, Liu (2019) used eight items to measure elementary students' collective self-efficacy, defined as students' shared beliefs in the collective power to achieve desired outcomes in the social-cognitive theory (Bandura, 2000).

The preceding instruments have been employed in diverse research designs to answer different research questions. Some studies aim to investigate students' average levels of self-efficacy beliefs or its subcategories (Dong, 2015) and compare them across gender (Li et al., 2012; Hu et al., 2010), grade levels (Chen & Bi, 2011; Hu et al., 2010; Li et al., 2012) and school types (Chen & Bi, 2011). Survey results indicated that high school students reported low self-efficacy beliefs in chemistry (Dong, 2015; Li et al., 2012) and physics (Hu et al., 2010). Empirical findings related to group differences in science self-efficacy beliefs are inconsistent, which may depend on the subject domain and student populations under investigation. Some studies also examined how science self-efficacy beliefs related to other variables including science achievement (Li et al., 2012; Hu et al., 2010) and the attribution of success and failure (Chen & Bi, 2011). Consistent with theoretical assumptions, high-performing students tend to report higher levels of perceived science self-efficacy beliefs than low-performing students (Li et al., 2012; Hu et al., 2010). Students' chemistry efficacy, including perceived learning abilities and perceived learning behaviors, mediates the effect of the attribution of success and failure on chemistry performance (Chen & Bi, 2011). The other two studies used the science self-efficacy instrument in pre- and post-experiment design to examine the influence of integrating innovative interventions in science instruction such as integrating immersive virtual reality into elementary science classrooms (Liu, 2019) or conducting problem-based learning activities in smart classrooms (Zhang et al., 2018).

Besides empirical studies, some journal articles are published in the format of theoretical papers that have explained definitions and functions of self-efficacy beliefs and how educational theories provide implications for teachers' science instruction. These suggested pedagogical strategies targeting students' self-efficacy beliefs include employing stratified and personalized instruction and assessment (Chen, 2012; Bao, Ding, Cheng, & Ma, 2012), enhancing group collaboration specifically between high-performing and low-performing students (Chen, 2012), providing opportunities for hands-on experiments and personalized support (Dong, 2015), and solving scientific problems in daily lives (Huang, 2020). Given the close association between self-efficacy beliefs and academic performance, science teachers may assist low-performing students in the following ways for improving their science self-efficacy beliefs: setting explicit and appropriate goals at different stages, attributing their success or failure in positive ways, and creating opportunities for perceiving success through dividing learning goals into small tasks and learning from role models about effective learning strategies (Bao et al., 2012; Dong, 2015; Gu, Yu, Xing, & Li, 2011; Liu & Zhang 2012; Man, Huang, & Li, 2016; Sun, 2013).

Achievement Emotions

Achievement emotions are defined as various emotions students experienced in academic settings including outcome-related emotions (e.g., shame, pride, and anxiety) and activity-related emotions (e.g., enjoyment, anger, and boredom) (Pekrun, 2006). Besides the aforementioned classification based on object focus, achievement emotions can also be divided into positive and negative emotions based on the valence or activating emotions and deactivating emotions based on the degree of activation. Compared with self-efficacy beliefs, the research history of achievement emotions is relatively short and only four studies have been identified in the past ten years. These studies examined students' achievement emotions in different disciplinary contexts. For example, Li, Wang, and Guo (2020) adapted items from Goetz et al.'s (2012) instrument and examined high school students' emotions during chemistry homework. Results showed that there were no gender differences in five emotions (i.e., enjoyment, pride, nervousness, anger, and boredom) while 11th-grade students reported higher negative emotions than 10th-grade and 12th-grade students. What is more, consistent with findings conducted in Western contexts (e.g., Dettmers, Trautwein, Lüdtke, Goetz, Frenzel, & Pekrun, 2011), positive emotions during chemistry homework were positively related to chemistry achievement while negative emotions showed the opposite pattern.

The other studies adapted items from Chinese Teenagers' Achievement Emotions Questionnaire (Dong & Yu, 2007) and developed instruments measuring students' achievement emotions in biology and physics respectively. In the biology domain, Cao et al. (2020) first examined eighth-grade and 11th-grade students' positive emotions across gender and grade levels. Then they investigated how students' perceived positive activating and deactivating emotions related to cognitive engagement. Results showed that there were no gender differences in positive activating emotions while positive deactivating emotions differed across gender. Female students perceived lower positive deactivating emotions than male students. Similarly, there were no significant differences either in positive activating or deactivating emotions across grade levels. However, the effect of positive emotions on cognitive engagement may differ across grade levels. Specifically, enjoyment, calmness, and satisfaction positively predicted eighth-grade students' perceived cognitive engagement while only calmness and satisfaction positively predicted 11th-grade students' perceived cognitive engagement.

Guo, Shen, and Yao (2011) also adapted items from Chinese Teenagers' Achievement Emotions Questionnaire (Dong & Yu, 2007) to measure students' achievement motions in the physics domain. They first invited undergraduate students who majored in psychology to revise items with the purpose of making the language translation and expression suitable for Chinese students. The confirmatory factor analysis was used to examine the survey's internal structure. The final instrument included 72 items measuring students' achievement emotions in four dimensions: 15 items measured positive activating emotions (i.e., pride, enjoyment, and hope), 14 items measured positive deactivating emotions (i.e., contentment, calmness, and relaxation), 17 items measured negative activating emotions (i.e., anxiety, shame, and anger), and 21 items measured negative deactivating emotions (i.e., boredom, helplessness, sadness, tiredness-upset). Survey results indicated main effects of gender and school type and interactions between demographic variables: male students had higher positive emotions and lower negative emotions than female students; students from top-ranked schools had higher positive emotions than ordinary-school students while there were no significant differences in negative emotions across school types. Regarding the interaction between gender and school type, male students from top-ranked schools had higher positive activating emotions than females and there were no gender differences in positive activating emotions for students from ordinary schools. Regarding the interaction between school type and grade level, students from top-ranked middle schools had higher positive deactivating emotions and lower negative deactivating emotions than students from top-ranked high schools. Students from ordinary middle schools had lower positive deactivating emotions than students from ordinary high schools. There were no significant differences in negative deactivating emotions between ordinary middle-school students and ordinary high-school students.

They further examined how achievement emotions related to outcome variables (i.e., physics achievement) and other motivational variables (i.e., environmental goal structure and goal orientation) (Guo, Shen, & Yao, 2012). Results showed that positive activating emotions and positive deactivating emotions were positively related to students' physics achievement while one specific negative activating emotion (i.e., shame) and three types of negative deactivating emotions (i.e., helplessness, boredom, and tiredness-upset) showed the opposite pattern. The environmental mastery goal structure positively predicted positive activating emotions, positive deactivating emotions, and negative activating emotions while the environmental performance goal structure positively predicted negative activating emotions and negative deactivating emotions. The individual student's achievement goal orientation mediated the effect of environmental goal structure on achievement emotions while achievement emotions mediated the relation between achievement goal orientation and physics achievement. The previously mentioned four empirical studies contribute to understanding students' achievement emotions in Chinese science classroom contexts.

Attitudes and Interests

Students' attitudes and interests are commonly used terms in science curriculum standards to describe students' motivations for learning science. However, there still exist debates on definitions or dimensions of these constructs. The recent literature has discussed the preceding two variables from theoretical and practical perspectives. Theoretically, researchers tried to identify how students' science attitudes or interests related to other variables (e.g., classroom environments and learning style preferences). For example, Zhou, Jiang, and Qin (2019) adapted items from a multidimensional questionnaire – the Test of Science-Related Attitudes (TOSRA; Fraser, 1981; Cohn & Fraser, 2016) – to measure students' attitudes and interests of attending chemistry classes. Results showed that students' attitudes toward chemistry were positively related to their views of classroom environments including student cohesiveness, task orientation, teacher support, involvement, cooperation, and equity. Zheng (2018) examined how students' interests in learning biology related to their perceptual learning style preferences (i.e., visual, auditory, tactile, and kinesthetic) across gender. The interest instrument was adapted from previous research (Zhang, 2005) and included four areas: motivation, learning style, interest in hands-on experiments, and biology teachers. Results showed that there were no gender differences in students' perceptual learning style preferences and biology interests. Tactile learners had the highest levels of motivation while auditory learners had the lowest. Visual and auditory learners had higher expectations for biology teachers while tactile learners were most interested in doing experiments. These findings suggest that students' science interests and attitudes are closely associated with internal personal characteristics (e.g., learning styles) and external classroom environments (e.g., teaching styles, classroom atmosphere). Therefore, other researchers aim to propose suggestions for improving students' science interests from a practical perspective. These proposed pedagogical strategies include organizing field trips and specimen exhibitions, designing classroom posters, and diversifying classroom activities (e.g., utilizing online resources or integrating inquiry-based activities) to enhance student involvement (Tan, 2016; Zhou, Jiang, & Qin, 2019).

Motivations

Motivation is another broad and prevalent term that describes the driving force directing individuals' actions and decisions. Considering its central role in the field of educational psychology, there have been numerous empirical studies using various instruments to examine students' motivations in past years. Similar to other constructs, existing research also discusses motivation from a theoretical perspective and a practical perspective. Some researchers have collected survey data to quantitatively measure and compare students' motivations for science learning across demographic variables and investigate their effect on academic performance while other researchers aim to provide suggestions for innovating teachers' pedagogical practices targeting enhancement of students' motivations for learning science.

Due to the conceptual vagueness, students' motivations for learning science were constructed and measured in diverse ways in empirical studies. For example, Liu and Zhang (2013) adapted survey items from Students' Motivation toward Science Learning Questionnaire (SMTSL) and measured motivations with 35 items in six categories: self-efficacy, active learning strategies, science learning value, performance goal, achievement goal, and learning environment stimulation (Tuan, Chin, & Shieh, 2005). The authors then further compared middle school students' science motivations across gender, grade levels, and school types. Results showed that six categories of motivations displayed different patterns of group differences: male students reported higher levels of self-efficacy, achievement goal, and learning environment stimulation than female students; students from urban schools reported higher levels of active learning strategies, science learning value, and learning environment stimulation than students from rural schools. In contrast, students from urban schools reported lower achievement goals than students from rural schools. What is more, students' motivation related to science textbooks and science teachers, measured by survey items developed by Hasan (1985), positively predicted their active learning strategies, science learning value, performance goal, and learning environment stimulation.

Besides the described approach of measuring motivations in six categories, other researchers tended to categorize this construct into intrinsic motivation (e.g., interest and curiosity) and extrinsic motivation (e.g., getting high grades and pursuing future careers). For example, Wu, Lu, and Wu (2019) examined how intrinsic motivation and extrinsic motivation of learning chemistry related to students' scientific inquiry abilities (e.g., asking questions, making hypotheses, planning, collecting evidence, drawing conclusions, providing explanations, evaluation, reflection, and communication). Results showed that both intrinsic motivation and extrinsic motivation positively predicted students' scientific inquiry abilities. Specifically, intrinsic motivation significantly predicted students' abilities of reflection, evaluation, and communication while extrinsic motivation significantly predicted students' abilities of making hypotheses. Li and Zhang (2019) adapted items from Chemistry Learning Motivation Questionnaire developed by Salta and Koulougliotis (2015) and examined how 11th-grade students' perceived motivation related to self-efficacy beliefs (i.e., learning abilities and learning behaviors) and chemistry achievement. The intrinsic motivation included the self-determination factor and the interest factor while the extrinsic motivation included the grade motivation and the career motivation. They found that high school students' perceived motivations and self-efficacy in learning chemistry were above the average estimate. Male students had higher intrinsic motivation and self-efficacy beliefs than female students. Chemistry self-efficacy beliefs mediated the relation between perceived motivation and chemistry achievement.

Intrinsic motivation and extrinsic motivation for learning are often described in different terms. Ausubel (1963) noted that the development of intrinsic motivation was accompanied by the acquisition of content knowledge. Enhancing the initiative in seeking knowledge for its own sake – cognitive drive – was the most effective strategy of developing students' intrinsic motivation. Simultaneously, extrinsic motivation was represented in the need for ego enhancement and career advancement, which contributed to obtaining positive encouragement from parents and teachers. Based on these theoretical hypotheses, some researchers measured motivations in such dimensions as cognitive drive, ego-enhancement drive, afflictive drive, and so on. For example, Liu and Ju (2013) revised existing items and compared how low-performing and high-performing high school students differed in cognitive drive, ego-enhancement drive, and afflictive drive. Results showed that both low-performing and high-performing students reported higher levels of cognitive drive than ego-enhancement drive and afflictive drive. What is more, the need for self-actualization was the main source of ego-enhancement drive for high-performing students while the pressure from college entrance examination was the main source of ego-enhancement drive for low-performing students. Hong (2013) examined eighth-grade students' motivations for learning physics in four dimensions: Cognitive drive, ego-enhancement drive, afflictive drive, and self-efficacy beliefs. Results indicated that the perceived ego-enhancement drive of learning physics was highest while the perceived self-efficacy was lowest. Male students had higher cognitive drive than female students but there were no significant gender differences in other variables. Yang (2013) incorporated different theoretical perspectives and measured ninth-grade students' motivations for learning chemistry in six areas: Cognitive drive, aspiration drive, goal drive, obligation drive, afflictive drive, and drive for avoiding social pressure. Findings regarding average levels of various types of drive and gender differences are not described here because the result section of this article only listed descriptive statistics and the lack of significant tests threatened the reliability of conclusions.

In addition, some studies focused on underrepresented student popula-tions and provided suggestions for science teachers' pedagogical strategies for improving students' motivations. For example, Zhou (2018) reported that minority students' motivations for learning physics were relatively low. Ben (2011) examined what internal or external factors resulted in negative motivations among female students ranging from eighth grade to 12th grade. Survey results showed that external rigid requirements and challenges, nega-tive feedback from others due to failures, and gender stereotypes in tradi-tional values negatively influenced female students' motivations for learning physics. Based on existing empirical findings, researchers suggested following pedagogical techniques for promoting students' motivations: Scaffolding problem-solving scenarios (Ben, 2011; Fu, 2011; Liu & Zhang, 2012), training students to attribute positively (Gu, Yu, Xing, & Li, 2011), creating a harmo-nious classroom atmosphere through balancing cooperation and competition, using rewards and punishments appropriately, connecting science content with daily lives (Wang & Ren, 2019), integrating historical stories and text-book pictures (Wu & Yang, 2013), creating opportunities of conducting in-class and extracurricular experiments (Fu, 2011; Wu & Yang, 2013; Wang & Ren, 2019), and improving the artistry and scientificity of the instructional language (Fu, 2011).

Summary and Discussion

The objective of the literature review section was to identify main themes and approaches in science education research that has focused on students' moti-vations and emotions in the past ten years. As the research on motivations has a long history and previous researchers have adopted various terms to describe this construct, it is difficult if not impossible to exhaustively identify and cover all related words in the corpus of literature. Murphy and Alexander (2000) have identified 20 terms relevant to academic motivation and achievement including goal (e.g., goal orientation, social goal), motivation (intrinsic moti-vation and extrinsic motivation), interest, and self-schema (e.g., attribution, self-efficacy). Based on this corpus of motivational terms and affective vari-ables selected in the previous literature reviews (e.g., Flaherty, 2020), I nar-rowed my attention to particular terms that are commonly used in Chinese language and literatures to describe students' motivations and emotions. These terms, including self-efficacy beliefs, achievement emotions, attitudes, interests, and motivations, were used as key words for my electronic data-base search.

The qualitative content analysis indicates that existing research is domi-nated by two types of published articles: Quantitative research and theoretical papers. These quantitative studies aim to examine single or multi-constructs of students' motivational or emotional perceptions while theoretical papers usually provide suggestions for improving science teachers' pedagogical strategies targeting students' motivations and emotions, which are hypothesized to have a big influence on students' learning behaviors and outcomes in academic settings. Regarding quantitative research, most empirical studies often adapted items from existing instruments (e.g., Dong, 2015; Guo et al., 2011; Li et al., 2020; Liu, 2019; Zhang et al., 2018) while only a few studies developed their own instruments (e.g., Hu et al., 2010; Li et al., 2012). Then survey data was collected from middle-school or high-school students to measure average levels of motivational or emotional variables and compare them across demographic information (e.g., gender, grade levels, and school types) and between different student groups (low-performing and high-performing students). Some studies examined intercorrelations between different motivational or emotional variables and their relationships with science performance (e.g., Cao et al., 2020; Chen & Bi, 2011; Guo et al., 2012). Two studies used these variables as indicators of examining the effectiveness of pedagogical interventions (e.g., Liu, 2019; Zhang et al., 2018). Survey results showed that students reported low to moderate levels of self-efficacy beliefs (Dong, 2015; Hu et al., 2010; Li et al., 2012; Li & Zhang, 2019). Findings on gender differences were somewhat inconsistent: Some studies reported that male students had higher levels of self-efficacy beliefs (Liu & Zhang, 2013), intrinsic motivations (Li & Zhang, 2019), cognitive drive (Hong, 2013), and positive emotions (Guo et al., 2011), whereas other studies reported no significant differences in specific types of positive or negative emotions (Cao et al., 2020; Li et al., 2020) and biology interests (Zheng, 2018) between male and female students. Findings drawn from comparisons of students at different grade levels, from different types of schools (urban versus rural and top-ranked versus ordinary), and with high or low performance showed similar patterns. There also existed interactions between demographic variables (e.g., Guo et al., 2011), which confirm important roles of situational or contextual factors in shaping students' motivational and emotional perceptions.

There were several possible explanations for the lack of coherence. First, similar to the Western psychological research community, different ways of conceptualizing or operationalizing motivational beliefs or emotions may lead to different research results (Gaspard et al., 2015; Hulleman, Schrager, Bodmann, & Harackiewicz, 2010; Marsh, Craven, Hinkley, & Debus, 2003; Murphy & Alexander, 2000). For example, some studies treated self-efficacy as one component of motivations (e.g., Hong, 2013; Liu & Zhang, 2013) while other studies treated them as separate and parallel constructs (e.g., Li & Zhang, 2019). Even for the same construct, researchers may define its subdimensions in different approaches. Some researchers measured motivations in two aspects (i.e., intrinsic motivation and extrinsic motivation) (e.g., Ding,

2014; Li & Zhang, 2019; Wu et al., 2019) while others measured motivations in more diverse dimensions (i.e., cognitive drive, ego-enhancement drive, and afflictive drive) (e.g., Hong, 2013; Liu & Ju, 2013). Therefore, differences in research findings may result from the variability in instruments used to measure the same construct. Second, different student samples may also contribute to the inconsistent pattern of findings. In the reviewed literature, participants ranged from fourth grade to 12th grade while schools ranged from rural schools to urban schools. There was also one study collecting data from students in minority autonomous regions (Zhou, 2018). As science teachers' pedagogical strategies or classroom learning activities heavily depend on their specific school contexts and student characteristics, such contextual factors influence students' motivations or emotions of learning science. What is more, as students' motivational beliefs and achievement emotions are domain-specific (Guo, Marsh, Parker, Morin, & Dicke, 2017; Pekrun, 2006), findings on gender differences are also closely associated with specific science subjects under investigation (e.g., physics, chemistry, or biology). Finally, there exist some methodological issues which may threaten the reliability of empirical findings. Though many studies adapted items from existing instruments, the lack of a detailed description of their procedures of translation or revision may fail to provide evidence for the instrument's cross-cultural validity. For example, many researchers only simply mentioned that they invited teachers or experts from different fields to provide feedback for improving the instrument's content validity before data collection and Cronbach α was commonly calculated to test its internal reliability. However, with very few exceptions, many of them did not conduct exploratory or confirmatory factor analysis to examine the instrument's internal structure. As Eastern cultural norms or values shape students' perceptions and expressions of motivations and emotions, factors emerged from statistical analysis might be different from theoretically hypothesized classifications. For example, compared with the division between positive emotions and negative emotions, Gong and Bergey (2020) found that Chinese students' achievement emotions in chemistry classes had two distinct factors: positive emotions and shame. Therefore, future research should take the role of Eastern cultures into consideration and improve the conceptualization clarity and measurement consistency of motivations and emotions (Hulleman et al., 2010).

Conclusions and Future Directions

The Confucian tradition treats teachers as the authority in classrooms. Consequently, traditional Chinese science education research in past years has focused on developing and implementing lesson plans that cultivate students' core literacies described in national science curriculum standards. Despite the fact that researchers and teachers agree about important roles of motivations and emotions in influencing students' learning behaviors and academic performance, the scarcity of related research indicates that this interdisciplinary field merits further investigation. To address this issue, this chapter first analyzed how elements of motivations and emotions were represented in science curriculum standards and then reviewed previous articles discussing or investigating students' motivations and emotions in the science domain during the past ten years. The goal is to provide insights for education researchers and practitioners into the interdisciplinary research between science education and education psychology. The qualitative content analysis showed that scientific attitude was the most prominent word that was related to motivations in national science curriculum standards. At present, schoolteachers are encouraged to adopt student-centered approaches for enhancing students' interests and perceived values of learning science, even though the great pressure of school examinations may hinder the progress of school reforms. Regarding the literature review, this study first identified 37 articles based on the inclusion criteria and then classified them by variables and methods used in these research studies. Findings showed that existing research mainly adopted quantitative approaches to examine students' self-efficacy beliefs, achievement emotions, attitudes, interests, and motivations and their relations with outcome variables, which provided implications for science teachers' pedagogical strategies aiming to improve students' learning experiences. However, the variability in conceptualizing and measuring these variables has led to inconsistent research findings. The reliability and validity of some measurement scales are undermined due to methodological issues. Therefore, I suggest that future research should take Eastern cultural contexts into consideration and provide more empirical evidence (e.g., factor analysis) for the cross-cultural validity of instruments. What is more, qualitative methods (e.g., interviews and open-response questions) or mixed methods can be used to examine students' motivational and emotional variables more deeply. For example, how do contextual factors (e.g., cultural values, classroom environments, and subject characteristics) or interventional strategies (e.g., teacher support) influence students' motivations and emotions? Why do some high-performing students still have low science self-efficacy beliefs? What are growth trajectories of students' motivations and emotions in the science domain? Only after fully understanding underlying mechanisms can researchers better transform theoretical findings into effective strategies of teaching science in practice. In addition, it should be noted that while this study only included some key terms related to motivations and emotions, future research could include other terms (e.g., goal orientation, agency and self-concept) to conduct literature review more comprehensively.

Appendix

Table A1 The reviewed literature related to motivations and emotions in the domain of science

No.	Author/Year	Domain/Context	Format/Data	Participants	Variables	Research Questions
Self-Efficacy Beliefs						
1	Xiao and Hu (2017)	Chemistry	Literature Review		Self-efficacy	
2	Chen and Bi (2011)	Chemistry	Survey	10th grade – 12th grade	Self-efficacy, attribution, and achievement	gender, grade level, performance, and school type
3	Dong (2015)	Chemistry	Self-developed Survey	12th grade	Self-efficacy	
4	Hu, Zhang, and Wu (2010)	Physics	Survey	10th grade – 12th grade	Self-efficacy	gender, grade level, and performance
5	Huang (2020)	Physics	Theoretical Paper			
6	Sun (2013)	Physics	Theoretical Paper	Low performing students		
7	Bao, Ding, Cheng, and Ma (2012)	Biology	Theoretical Paper	Low performing students		
8	Chen (2012)	Physics	Theoretical Paper			
9	Li, Fan, and Tao (2012)	Chemistry	Survey	10th grade – 12th grade	Self-efficacy	gender, grade level, and performance
10	Zhang, Dong, Bai, Xiong, and Zhu (2018)	Science/problem-based learning in smart classrooms	Experiment design: survey	4th grade	Self-efficacy of learners for inquiry engagement, interest, cognitive load, inquiry	Self-efficacy: compare cognitive load, inquiry
11	Liu (2019)	Science/Immersive virtual reality	Experiment design: survey and interviews	4th grade	Individual and collective self-efficacy	Traditional group and experiment group

(Continued)

Table A1 (Continued)

No.	Author/Year	Domain/Context	Formal/Data	Participants	Variables	Research Questions
Achievement Emotions						
12	Li, Wang, and Guo (2020)	Chemistry	Survey	10th grade – 12th grade	Homework-related achievement emotions	Grade level and performance
13	Guo, Shen, and Yao (2012)	Physics	Survey	8th grade – 12th grade	achievement emotions, achievement goal, environmental goal structure	
14	Cao, Ying, Cao, Dai, and Yao (2020)	Biology	Survey	8th grade, 11th grade	achievement emotions, engagement	gender, grade level
15	Guo, Shen, and Yao (2011)	Physics	Validated Survey	8th grade, 10th grade, 11th grade	achievement emotions	gender, grade level, school type, achievement
Attitude and Interest						
16	Zhou, Jiang, and Qin (2019)	Chemistry	Survey	10th grade, 11th grade	Attitude, classroom environment	
17	Man, Huang, and Li (2016)	Physics	Theoretical Paper		Motivation, self-efficacy, attribution	
18	Tan (2016)	Biology	Theoretical Paper		Interest	
19	Peng and Huang (2020)	Biology	Literature Review		Interest, learning strategy, motivations	School type
20	Zheng (2018)	Biology	Survey	10th grade	Interest (i.e., motivation, learning style, interest in experiments, and teachers), learning style preference	Gender

Motivation

#	Reference	Subject	Method	Grade	Constructs	Moderators
21	Deng, Wang, and Ye (2019)	Chemistry	Survey	10th grade	Prior knowledge, motivation, attitude, metacognition, self-regulation	
22	Liu and Zhang (2013)	Science	Survey	7th–9th grade	Motivation (i.e., self-efficacy, active learning strategy, science learning value, performance goal, achievement goal, learning environment stimulation)	gender, grade level, school type, teacher, textbook
23	Yang (2013)	Chemistry	Survey	9th grade	Motivation	
24	Ding (2014)	Chemistry	Survey	11th grade	Motivation (intrinsic and extrinsic)	
25	Liu and Ju (2013)	Chemistry	Survey	10th grade and 11th grade	Motivation (cognitive drive, ego-enhancement drive, afflictive drive)	Achievement level
26	Li and Zhang (2019)	Chemistry	Survey	11th grade	Motivation (intrinsic and extrinsic), self-efficacy (ability and behavior)	
27	Zhu and Qin (2013)	Chemistry	Literature Review		Motivation	
28	Liu and Zhang (2012)	Chemistry	Theoretical Paper		Motivation	
29	Wu, Lu, and Wu (2019)	Chemistry	Survey	10th grade	Scientific inquiry and motivation (intrinsic and extrinsic)	
30	Wang and Ren (2019)	Physics	Theoretical Paper		Motivation	

(Continued)

Table A1 (Continued)

No.	Author/Year	Domain/Context	Format/Data	Participants	Variables	Research Questions
31	Xiang, Yao, Guo, and Fortes (2020)	Physics	Literature Review		Motivation	
32	Ben (2011)	Physics	Survey	female students from 8th-grade to 12th grade	Motivation	
33	Zhou (2018)	Physics	Survey	Middle- and high-school minority students	Motivation	Gender
34	Hong (2013)	Physics	Survey	8th grade	Motivation (cognitive drive, ego-enhancement drive, afflictive drive, self-efficacy)	
35	Gu, Yu, Xing, and Li (2011)	Biology	Theoretical Paper		Motivation and attribution theory	
36	Wu and Yang (2013)	Physics	Theoretical Paper		Motivation, interest, emotion, and volition	
37	Fu (2011)	Biology	Theoretical Paper		Interest and motivation	

References

Abd-El-Khalick, F., Boujaoude, S., Duschl, R., Lederman, N. G., Mamlok-Naaman, R., Hofstein, A., ... Tuan, H. L. (2004). Inquiry in science education: International perspectives. *Science Education*, *88*(3), 397–419.

Ashby, F. G., Isen, A. M., & Turken, A. U. (1999). A neuropsychological theory of positive affect and its influence on cognition. *Psychological Review*, *106*(3), 529–550. https://doi.org/10.1037/0033-295X.106.3.529

Ausubel, D. P. (1963). A teaching strategy for culturally deprived pupils: Cognitive and motivational considerations. *The School Review*, *71*(4), 454–463.

Baker, D., & Taylor, P. C. (1995). The effect of culture on the learning of science in non-western countries: The results of an integrated research review. *International Journal of Science Education*, *17*(6), 695–704.

Bandura, A. (1986). The explanatory and predictive scope of self-efficacy theory. *Journal of Social and Clinical Psychology*, *4*, 359–373. doi:10.1521/jscp.1986.4.3.359

Bandura, A. (2000). Exercise of human agency through collective efficacy. *Current Directions in Psychological Science*, *9*(3), 75–78.

Bao, Z., Ding, C., Cheng, Y., & Ma, L. (2012). Cultivating underachieving students' self-efficacy beliefs in biology. *Ability and Wisdom*, *22*, 56. (In Chinese)

Ben, G. (2011). Investigating female students' negative motivations for learning physics. *Physics Teacher*, *32*(7), 7+11. (In Chinese)

Bian, Y. (2003). *The compilation and application of perceived academic self-efficacy scale*. [Doctoral dissertation, East China Normal University]. China National Knowledge Infrastructure (CNKI). (In Chinese)

Cao, B., Ying, T., Cao, L., Dai, M., & Yao, B. (2020). Investigating the relationship between students' positive emotions and cognitive engagement in biology. *Journal of Lanzhou Institute of Education*, *36*(6), 89–92. (In Chinese)

Chen, P. (2012). Using stratified activity design to improve middle school students' physics self-efficacy beliefs. *Physics Teacher*, *33*(07), 25–26. (In Chinese)

Chen, Y., & Bi, H. (2011). Students' chemistry self-efficacy beliefs and attributions of success and failure. *Chinese Journal of Chemical Education*, *32*(8), 23–27.

Cohn, S. T., & Fraser, B. J. (2016). Effectiveness of student response systems in terms of learning environment, attitudes and achievement. *Learning Environments Research*, *19*, 153–167.

Deng, C., Wang, X., & Ye, F. (2019). Investigating how students at different performance levels learn chemistry. *Journal of Huanggang Normal University*, *39*(3), 89–93. (In Chinese)

Dettmers, S., Trautwein, U., Lüdtke, O., Goetz, T., Frenzel, A. C., & Pekrun, R. (2011). Students' emotions during homework in mathematics: Testing a theoretical model of antecedents and achievement outcomes. *Contemporary Educational Psychology*, *36*(1), 25–35.

Ding, S. (2014). Investigating 10th-grade students' motivations of learning chemistry. *Ability and Wisdom*, *18*, 58. (In Chinese)

Dong, X. (2015). Investigating 12th-grade students' chemistry self-efficacy beliefs. *Chinese Journal of Chemical Education*, *36*(13), 55–58. (In Chinese)

Dong, Y., & Yu, G. (2007). The development and application of Chinese teenagers' achievement emotions. *Acta Psychologica Sinica*, *5*, 852–860. (In Chinese)

Eccles J. S. (2005). Subjective task value and the Eccles et al. model of achievement-related choices. In A. J. Elliot & C. S. Dweck (Eds.), *Handbook of competence and motivation* (pp. 105–121). London: Guildford Press.

Flaherty, A. A. (2020). A review of affective chemistry education research and its implications for future research. *Chemistry Education Research and Practice, 21*(3), 698–713. https://doi.org/10.1039/c9rp00200f

Fraenkel, J. R., Wallen, N. E., & Hyun, H. H. (2015). *How to design and evaluate research in education* (9th ed.). New York, NY: McGraw-Hill Education.

Fraser, B. J. (1981). *Test of Science-related attitudes handbook.* Melbourne, Victoria: Australian Council for Educational Research.

Fu, P. (2011). Cultivating middle school students' motivations and interests in learning chemistry. *Education and Teaching Forum, 28,* 131–132. (In Chinese)

Gaspard, H., Dicke, A. L., Flunger, B., Schreier, B., Häfner, I., Trautwein, U., & Nagengast, B. (2015). More value through greater differentiation: Gender differences in value beliefs about math. *Journal of Educational Psychology, 107*(3), 663–677.

Goetz, T., Nett, U. E., Martiny, S. E., Hall, N. C., Pekrun, R., Dettmers, S., & Trautwein, U. (2012). Students' emotions during homework: Structures, self-concept antecedents, and achievement outcomes. *Learning and Individual Differences, 22*(2), 225–234.

Gong, X., & Bergey, B. W. (2020). The dimensions and functions of students' achievement emotions in Chinese chemistry classrooms. *International Journal of Science Education, 42*(5), 835–856.

Gu, Y., Yu, J., Xing, G., & Li, Z. (2011). Stimulating learning motivation of middle school students in biology curriculum by means of attribution theory. *Journal of Hubei Normal University (Natural Science), 31*(4), 99–102.

Guo, J., Marsh, H. W., Parker, P. D., Morin, A. J. S., & Dicke, T. (2017). Extending expectancy-value theory predictions of achievement and aspirations in science: Dimensional comparison processes and expectancy-by-value interactions. *Learning and Instruction, 49,* 81–91. https://doi.org/10.1016/j.learninstruc.2016.12.007

Guo, L., Shen, J., & Yao, H. (2011). The exploration and application of physics academic emotions questionnaire revised for high school students. *Studies of Psychology and Behavior, 9*(4), 297–303. (In Chinese)

Guo, L., Shen, J., & Yao, H. (2012). The construction of physics academic emotion model based on the control-value theory. *Psychological Exploration, 32*(2), 153–159. (In Chinese)

Hasan, O. E. (1985). An investigation into factors affecting attitudes toward Science of secondary school students in Jordan. *Science Education, 69*(1), 3–18.

Hong, X. (2013). 8th-grade students' motivations for learning physics in the city of Shihezi in Xinjiang. *Journal of Bingtuan Education Institute, 23*(2), 74–76. (In Chinese)

Hu, X., Zhang, K., & Wu, J. (2010). The characteristics of senior middle school students' physics self-efficacy: Taking the students in one senior middle school in Shandong province as an example. *Curriculum, Teaching Materials and Method, 30*(8), 74–78. (In Chinese)

Huang, Y. (2020). Research on high school students' self-efficacy beliefs in physics. *Science & Technology Information, 18*(15), 124–125. (In Chinese)

Hulleman, C. S., Schrager, S. M., Bodmann, S. M., & Harackiewicz, J. M. (2010). A meta-analytic review of achievement goal measures: Different labels for the same constructs or different constructs with similar labels? *Psychological Bulletin, 136*(3), 422–449.

Ibrahim, A., Aulls, M. W., & Shore, B. M. (2016). Development, validation, and factorial comparison of the McGill Self-Efficacy of Learners for Inquiry Engagement (McSELFIE) survey in natural science disciplines. *International Journal of Science Education, 38*(16), 2450–2476.

Kwon, H., Yoon, K. L., Joormann, J., & Kwon, J. H. (2013). Cultural and gender differences in emotion regulation: Relation to depression. *Cognition and Emotion*, *27*(5), 769–782. https://doi.org/10.1080/02699931.2013.792244

Li, K., & Zhang, W. (2019). Relationships among high school students' learning motivation, learning self-efficacy and academic performance in chemistry. *Chinese Journal of Chemical Education*, *40*(21), 48–55. (In Chinese)

Li, N., Wang, H., & Guo, X. (2020). Investigating high school students' achievement emotions in doing chemistry homework. *Chemistry Teaching and Learning*, *518*(2), 90–93. (In Chinese)

Li, Y., Fan, Y., & Tao, R. (2012). High school students' chemistry self-efficacy beliefs in Kaifeng in the context of curriculum reform. *Chinese Journal of Chemical Education*, *33*(2), 57–59+63. (In Chinese)

Liu, J., & Zhang, Y. (2013). The characteristics and influencing factors of middle school students' science motivations. *Journal of Shanghai Educational Research*, *311*(5), 43–47. (In Chinese)

Liu, R. (2019). Empirical study of the effect of immersive virtual environment on learners' self-efficacy and collective efficacy: A case on the fourth-grade primary science curriculum. *Modern Educational Technology*, *29*(8), 72–78. (In Chinese)

Liu, Y., & Ju, X. (2013). Investigating low-performing students' motivations for learning chemistry in high schools. *Chinese Journal of Chemical Education*, *34*(11), 49–52. (In Chinese)

Liu, Y., & Zhang, C. (2012). Effective methods of inspiring students' learning motivation in chemistry teaching. *Science & Technology Information*, *417*(25), 121–122. (In Chinese)

Ma, L. (1999). *Knowing and teaching elementary mathematics: Teachers' understanding of fundamental mathematics in China and the United States*. Mahwah, NJ: Lawrence Erlbaum Associates.

Man, J., Huang, G., & Li, Y. (2016). Factors influencing high school students' physics achievement and related suggestions. *Asia-Pacific Education*, *56*(17), 142–143. (In Chinese)

Marsh, H. W., Craven, R. G., Hinkley, J. W., & Debus, R. L. (2003). Evaluation of the big-two-factor theory of academic motivation orientations: An evaluation of jingle-jangle fallacies. *Multivariate Behavioral Research*, *38*(2), 189–224.

Murphy, P. K., & Alexander, P. A. (2000). A motivated exploration of motivation terminology. *Contemporary Educational Psychology*, *25*(1), 3–53.

Nagy, G., Trautwein, U., Baumert, J., Köller, O., & Garrett, J. (2006). Gender and course selection in upper secondary education: Effects of academic self-concept and intrinsic value. *Educational Research and Evaluation*, *12*(4), 323–345.

Ogawa, M. (1986). Toward a new rationale of science education in a non-western society. *European Journal of Science Education*, *8*(2), 113–119. https://doi.org/10.1080/0140528860800201

Pajares, F., & Miller, M. D. (1994). Role of self-efficacy and self-concept beliefs in mathematical problem solving: A path analysis. *Journal of Educational Psychology*, *86*, 193. https://doi.org/10.1037//0022–0663.86.2.193

Pekrun, R. (2006). The control-value theory of achievement emotions: Assumptions, corollaries, and implications for educational research and practice. *Educational Psychology Review*, *18*(4), 315–341.

Peng, L., & Huang, S. (2020). The bibliometric analysis of students' psychological research on biology learning. *Biology Teaching*, *45*(12), 9–11. (In Chinese)

Pintrich, P. R., Smith, D., Garcia, T., & McKeachie, W. J. (1991). *A manual for the use of the Motivated Strategies for Learning Questionnaire (MSLQ)*. Ann Arbor, MI: University of Michigan, National Center for Research to Improve Postsecondary Teaching and Learning.

Salta, K., & Koulougliotis, D. (2015). Assessing motivation to learn chemistry: adaptation and validation of science motivation questionnaire ii with Greek secondary school students. *Chemistry Education Research and Practice, 16*(2), 237–250.

Schwarzer, R., & Jerusalem, M. (1997). Generalized self-efficacy scale. In J. Weinman, S. Wright, & M. Johnston (Eds.), *Measures in health psychology: A user's portfolio* (pp. 35–37). Windsor, UK: NFER-Nelson.

Skinner, E. A., & Chapman, M. (1984). Control beliefs from an action perspective. *Human Development, 27,* 129–133.

Skinner, E. A., Chapman, M., & Baltes, P. B. (1988). Control, means-ends, and agency beliefs: A new conceptualization and its measurement during childhood. *Journal of Personality and Social Psychology, 54*(1), 117.

Sun, S. (2013). Promoting underachieving students' self-efficacy beliefs in physics. *Popular Science (Science Education), 715*(1), 4. (In Chinese)

Tan, W. (2016). The discussion about how to incite high school students' interests in learning biology. *Chemical Management, 429*(32), 73. (In Chinese)

The Ministry of Education of the People's Republic of China. (2020a). *Revised High School Chemistry Curriculum Standards*. Beijing: People's Education Press. (In Chinese)

The Ministry of Education of the People's Republic of China. (2020b). *Revised High School Biology Curriculum Standards*. Beijing: People's Education Press. (In Chinese)

The Ministry of Education of the People's Republic of China. (2020c). *Revised High School Physics Curriculum Standards*. Beijing: People's Education Press. (In Chinese)

The Ministry of Education of the People's Republic of China. (2022a). *Chemistry Curriculum Standards of compulsory education*. Beijing: Beijing Normal University Press. (In Chinese)

The Ministry of Education of the People's Republic of China. (2022b). *Biology Curriculum Standards of compulsory education*. Beijing: Beijing Normal University Press. (In Chinese)

The Ministry of Education of the People's Republic of China. (2022c). *Physics Curriculum Standards of compulsory education*. Beijing: Beijing Normal University Press. (In Chinese)

The Ministry of Education of the People's Republic of China. (2022d). *Science Curriculum Standards of compulsory education*. Beijing: Beijing Normal University Press. (In Chinese)

Tuan, H. L., Chin, C. C., & Shieh, S. H. (2005). The development of a questionnaire to measure students' motivation towards science learning. *International Journal of Science Education, 27*(6), 639–654.

Wang, Q., & Ren, X. (2019). Research on motivations of physics learning based on the hierarchy of needs theory. *Education Modernization, 6*(74), 124–126. (In Chinese)

Wang, S. L., & Lin, S. S. (2007). The effects of group composition of self-efficacy and collective efficacy on computer-supported collaborative learning. *Computers in Human Behavior, 23*(5), 2256–2268.

Wang, W., Wang, J., Zhang, G., Lang, Y., & Mayer, V. J. (1996). Science education in the People's Republic of China. *Science Education, 80*(2), 203–222.

Weiner, B. (1986). *An attributional theory of motivation and emotion.* New York: Springer Verlag.

Wu, H., Lu, Q., & Wu, Y. (2019). The relationship between students' motivations for learning chemistry and their scientific inquiry abilities. *Journal of Teaching and Management, 766*(9), 31–34. (In Chinese)

Wu, X., & Yang, H. (2013). Strategies for cultivating high school students' non-intelligence factors in Physics. *Journal of Zunyi Normal College, 15*(2), 117–119. (In Chinese)

Xiang, Y., Yao, J., Guo, Y., & Fortes, D. (2020). The research on students' learning motivations in one specific subject: Using the physics chemistry as the example. *Physics Teaching, 42*(11), 4–8. (In Chinese)

Xiao, W., & Hu, Z. (2017). Research status and future directions of self-efficacy in high school chemistry education. *Chinese Journal of Chemical Education, 38*(23), 68–74. (In Chinese)

Yang, G. (2013). Investigating middle school students' motivations for learning chemistry. *Education and Teaching Forum, 125*(44), 174–175. (In Chinese)

Zhang, H. (2005). *The investigation and comparison of students' learning interests in biology.* [Doctoral dissertation, Tianjin Normal University]. China National Knowledge Infrastructure (CNKI). (In Chinese)

Zhang, Y., Dong, X., Bai, Q., Xiong, Y., & Zhu, Y. (2018). A study on inquiry engagement of students in smart classrooms: "Food travel in the body" as the case. *E-Education Research, 39*(5), 86–92. (In Chinese)

Zheng, S. (2018). Study on the relationship between biological learning interest and perceptual learning styles among 10th-grade students in Guangzhou. *Educational Measurement and Evaluation, 204*(1), 54–61. (In Chinese)

Zhou, G. (2018). An Investigation on the learning motivation of minority students in Physics. *The Guide of Science & Education, 343*(19), 183–185. (In Chinese)

Zhou, J., Jiang, M., & Qin, C. (2019). The relationship between high school students' learning attitudes and views of classroom environments in chemistry. *Education in Chemistry, 390*(9), 28–33. (In Chinese)

Zhu, M., & Qin, H. (2013). The literature review of research on motivations of learning chemistry in ten years. *Education in Chemistry, 10*(4), 13–16. (In Chinese)

Part 2

Students' Motivations and Emotions in Regular Science Classrooms

2 The Dimensions and Functions of Students' Achievement Emotions in Chinese Chemistry Classrooms

Introduction

Existing research in Western contexts has shown that achievement emotions are multifaceted and linked with achievement-related variables such as self-efficacy (Marchand & Gutierrez, 2012; Putwain, Sander, & Larkin, 2013) and engagement (Reschly, Huebner, Appleton, & Antaramian, 2010). Given that culture substantially shapes how students experience and express emotions through the values and norms that are emphasized in classroom contexts (Tsai, Knutson, & Fung, 2006), an open question is whether similar dimensions and relations are found in Chinese academic contexts. In the current study, I address this gap by examining achievement emotions in Chinese high school chemistry classrooms. Specifically, I investigate the dimensions of achievement emotions and whether these emotions mediate the relationship between self-efficacy and engagement. A better understanding of Chinese students' emotions can increase knowledge about the role of culture in regulating students' class-related emotions and further inform efforts to develop emotionally supportive classroom learning environments across cultural contexts.

In this paper, I first present an overview of control-value theory, which defines achievement emotions and theorizes relationships with self-efficacy and engagement, and briefly review empirical findings from studies conducted in Western contexts. I then discuss how cultural values and contexts influence students' perceptions of achievement emotions, self-efficacy beliefs, and engagement. Finally, I follow with a description of the current study.

Theoretical Framework: Achievement Emotions and Control-Value Theory

Achievement emotions, defined as emotions directly related to academic activities and outcomes (Pekrun, 2006), are critical predictors of students' academic performance and career choices (Schutz & Pekrun, 2007). Enjoyment, hope, pride, relief, anxiety, shame, hopelessness, anger, frustration, and boredom are commonly occurring emotions in academic settings (Pekrun, Goetz, Titz, & Perry, 2002a). As shown in Table 2.1, achievement

DOI: 10.4324/9781032649665-4

Table 2.1 A three-dimensional taxonomy of achievement emotions (Pekrun, 2006)

Object Focus	Positive		Negative	
	Activating	Deactivating	Activating	Deactivating
Activity Focus	Enjoyment	Relaxation	Anger Frustration	Boredom
Outcome Focus	Joy Hope Pride Gratitude	Contentment Relief	Anxiety Shame Anger	Sadness Disappointment Hopelessness

emotions can be characterized according to three dimensions: the degree of activation, valence, and object focus. With regard to *degree of activation*, achievement emotions are grouped into activating emotions (e.g., joy, frustration) and deactivating emotions (e.g., relief, sadness). In regard to *valence*, achievement emotions can be positive (e.g., gratitude, pride) and negative (e.g., shame, anger). In regard to *object focus*, achievement emotions can be classified as activity related (e.g., enjoyment, anger) and outcome related (e.g., pride, anxiety). As students are likely to experience various emotions in academic settings, these dimensions can be used for analyzing interrelations of different achievement emotions. For example, Pekrun et al. (2002a) reported that four clusters emerged with samples of Western university students: (a) enjoyment, hope, and pride; (b) relief; (c) anxiety, shame, and hopelessness; (d) anger and boredom. Emotions in the first and second clusters were positive activating and positive deactivating respectively while the third and fourth clusters represented negative emotions with mixed degrees of activation.

Control-value theory provides an integrative framework for examining how achievement emotions affect motivation and posits two key assumptions about the antecedent and effect of achievement emotions (Pekrun, 2006). The first assumption is that control-related beliefs predict students' perceptions of achievement emotions. Students perceive enjoyment when feeling in high control of achievement activities and frustration when feeling out of control. While control-related beliefs are related to multiple motivational constructs (Wigfield & Cambria, 2010), in the current study I focus on self-efficacy – an individual's judgments of his or her capability to organize and execute courses of action required to perform a context-related task successfully (Bandura, 1986; Pajares & Miller, 1994). The situation-specific nature of self-efficacy beliefs (Bandura, 2006; Linnenbrink & Pintrich, 2003) make it an appropriate construct for evaluating control-related beliefs within control-value theory (Luo, Ng, Lee, & Aye, 2016; Putwain & Symes, 2014). Students with higher self-efficacy beliefs are expected to appraise a situation as manageable, perceive a higher likelihood of success, and thus maintain positive emotions (e.g., enjoyment). In contrast, students with low self-efficacy beliefs are more likely to perceive a situation as a threat and the anticipation of failure increases negative emotions (e.g., anxiety) (Bandura, 1997; Zimmerman, 2000).

Prior research in Western contexts has supported the theorized relations among control-related beliefs and achievement emotions. Pekrun et al. (2004) examined German undergraduate students' achievement emotions in situations of taking exams and tests. They found that academic self-efficacy was positively related to positive emotions (i.e., joy, hope, and pride) and negatively related to negative emotions (i.e., anger, anxiety, shame, and hopelessness). Marchand and Gutierrez (2012) reported that perceived self-efficacy of American graduate students was a negative predictor for frustration and anxiety. In a sample of first-year undergraduate students in the United Kingdom, Putwain et al. (2013) identified that academic self-efficacy about one's study-related skills and behaviors was positively associated with pleasant emotions and negatively associated with negative emotions.

The second assumption in control-value theory is about the effect of achievement emotions on student engagement. For the current study, I define student engagement as effort, attention, and persistence during the initiation and execution of learning activities (Skinner, Furrer, Marchand, & Kindermann, 2008). Student engagement is presumed to be malleable and is regarded as a proximal indicator of students' academic retention, achievement, and resilience (Skinner et al., 2008). As achievement emotions are multifaceted, the interplay between valance and activation can produce four basic categories of emotions (i.e., positive activating, positive deactivating, negative activating, and negative deactivating) and exert different effects on student engagement. Positive activating emotions (e.g., enjoyment, hope) are reported to be positively related to students' self-reported effort, whereas negative deactivating emotions (e.g., boredom and hopelessness) show the opposite pattern (Pekrun et al., 2002a; Pekrun, Goetz, Titz, & Perry, 2002b; Pekrun, Goetz, Frenzel, Barchfeld, & Perry, 2011). The linkage between negative activating emotions (e.g., anger, anxiety, and shame) or positive deactivating emotions with engagement has received less attention by researchers, yet these relations provide a fuller understanding of how different emotions function in academic settings. Despite overall negative correlations between negative activating emotions and student engagement (Pekrun et al., 2004, 2011), there exist exceptional situations. For example, perceived anxiety and shame might stimulate students to invest more effort (e.g., reevaluate and modify learning strategies) to avoid failure (Pekrun & Linnenbrink-Garcia, 2012; Turner & Schallert, 2001). Based on control-value theory, achievement emotions mediate the relationship between self-efficacy beliefs and student engagement, such that a sense of control over outcomes influences emotions that in turn influence engagement. Self-efficacy may also directly affect student engagement by regulating the quantity of expended effort and the willingness to persist in tasks (Bandura, 1997). Students with high self-efficacy beliefs are expected to persist and spend more effort in the face of difficulty while students with low self-efficacy beliefs are more likely to doubt themselves and give up easily when confronting challenges (Linnenbrink & Pintrich, 2003; Usher & Pajares, 2008). Existing studies have examined how self-efficacy predicts student engagement

(Schunk, 1991, 2001). For example, Walker, Greene, and Mansell (2006) found that American undergraduate students' self-efficacy uniquely predicted meaningful cognitive engagement after controlling for other variables such as intrinsic motivations and academic identification. Therefore, self-efficacy may directly shape student engagement, outside of its effect on achievement emotions.

Control-value theory also postulates that students' emotional experiences are situated within specific cultural contexts and that features of these contexts shape how emotions operate. In the following section, I review how different cultures regulated students' expressions or perceptions of three variables and presented empirical findings conducted in Eastern contexts.

Cultural Influence on Achievement Emotions, Self-Efficacy, and Engagement

Achievement Emotions

Cultural values and norms can influence how individuals express, experience, and interpret emotions (Markus & Kitayama, 1991; Zembylas, 2003). Chinese cultural characteristics and values are deeply rooted in Confucian traditions, which stress order, structure, and respect for authority. In cultures shaped by Confucian traditions, academic success is an important pathway for individuals to seek recognition for themselves and their family. For example, admission to prestigious universities is not only self-advancement but also the fulfillment of parental expectations, which brings glory and face to the community. Face, defined as "the confidence of society in the integrity of ego's moral character" (Hu, 1944, p. 45), is closely associated with one's sense of dignity and reputation in Chinese society. Whereas academic success is a badge of honor for both the individual and community, poor academic performance or expulsion from school leads to feelings of shame and loss of family face (Gow, Balla, Kember, & Hau, 1996). Such outcomes have direct and detrimental consequences for one's interactions with others in the community (Hu, 1944).

In addition to the influence of Confucian cultures, some scholars have argued that individualistic Western and collectivistic Eastern cultures have distinct concepts of the student self (Oyserman, Coon, & Kemmelmeier, 2002), with implications for achievement emotions. In individualistic cultures (e.g., Western classrooms), the student self is perceived to be independent and unique from others. Students are often encouraged to pursue individual goals and express personal feelings openly and directly (Oyserman et al., 2002). By contrast, the student self in collectivistic cultures (e.g., Eastern cultures) is perceived to be interdependent and cannot be separated from others. Students are expected to fulfill their social obligations, develop strong group identities, maintain harmony with others, and support the goals of others with whom they share valued social relationships (Eid & Diener, 2001). Thus, unique social norms, expectations, and self-perceptions in different cultural contexts can prompt an individual to express achievement emotions in different ways. While students in individualistic cultures are more likely to express various emotions as independent selves, students in collectivistic and Confucian cultures are more likely to express emotions that reflect the values of hierarchical social relationships and group harmony (Hsu, 1971). Accordingly, in Eastern cultural contexts students may be more likely to express culturally desirable emotions, such as low-activating emotions (e.g., calmness), and to constrain undesirable feelings, such as high-activating emotions (e.g., anger) (Wong & Tsai, 2007). Mesquita and Karasawa (2002) investigated cultural differences in undergraduate students' everyday emotions by administering emotion questionnaires four times a day for one week. They reported that Asian students more often experienced no emotions than American peers. Similarly, other researchers have noted that East Asians are less likely to express or even notice their emotions than North Americans (Heine, 2001). The pattern of restraining emotions potentially reflects cultural values that emphasize hierarchy, harmony, and filial piety in Chinese society.

Chinese cultural influences may shape the importance and function of one negative and activating emotion in particular: shame. The importance of shame has been elaborated in Chinese cultures (Li, Wang, & Fischer, 2004). For example, there is an extensive vocabulary for the perception of different types of shame (Russell & Yik, 1996). Wilson (1981) defined a verbal scale of shame in Chinese, ranging from the least to the most intense feelings: unease or shyness (害羞), embarrassment (不好意思, 尴尬), losing face (丢脸), deep shame (惭愧), and extreme shame (无耻, 不要脸). Individuals who are labelled as extreme shame might be humiliated, alienated, or even isolated from others and thus lose personal powers or values. The incitement of shame, which is usually associated with individuals' self-evaluation of failing to meet specific standards, can serve to socially control individuals in Asian countries by emphasizing personal responsibilities and accountability to meet social expectations (Marsella, Murray, & Golden, 1974). Within classroom environments, over the past few decades, Chinese teachers may announce each student's grades and class rankings after school examinations to incite the feeling of shame for those who have fallen behind. But in recent years, school administrators and teachers are suggested to protect student privacy and keep these information confidential. The prominence and importance of shame in Chinese contexts is also hinted at in previous studies that have found Chinese high school students reported more shame but less anger in mathematics than German peers (Frenzel, Thrash, Pekrun, & Goetz, 2007).

Self-Efficacy and Engagement

Cultural influences also shape teachers' and students' expectations about the right way of expressing self-efficacy and engagement in classrooms. Given that the construct of the self is more tied to a collective rather than an individualistic identity, students in Eastern cultures are often encouraged to be modest and make self-effacing responses (Bond, Leung, & Wan, 1982). Such behaviors are consistent with the Chinese proverb: "modesty helps one to go forward, whereas conceit makes one lag behind". As a result, Asian students tend to underestimate their abilities (Zusho & Pintrich, 2003). Previous research has compared students' general or math self-efficacy at different age levels across cultural contexts and found that students in Eastern cultures reported lower levels of self-efficacy than peers in Western cultures (Klassen, 2004; Lee, 2009; Scholz, Doña, Sud, & Schwarzer, 2002).

Similarly, cultural contexts influence what it means to be engaged and the opportunities classroom structures and instructional approaches provide for different forms of engagement. Chinese high school class sizes—which commonly consist of 50 or more students—are typically large compared to some Western countries (e.g., United States). Accordingly, Chinese teachers typically use whole-class and teacher-directed instruction approaches, whereas small-group, individual, or constructivist instructional approaches are more common in US classrooms (Stigler & Perry, 1988). In such settings, Chinese students have few opportunities to interact with other students during instruction; instead, they rely heavily on teachers' explanations and students seldom interrupt lectures to ask questions, which may take away time for the whole class. As a result, student engagement in Chinese classrooms is characterized by maintaining attention toward teachers' directions and explanations. Chinese teachers' perceptions of students' most frequent troublesome misbehaviors are daydreaming, being inattentive, and not answering questions. By contrast, in Western context, where learning environments are more participatory, where student questions are encouraged, and students have opportunities to cooperate with classmates in small-group activities, teachers report talking out of turn as a primary misbehavior (Ding, Li, Li, & Kulm, 2008). The differences in classroom environments (independence versus collaboration) potentially influence students' perceptions of being engaged.

In summary, as Eastern cultural features are likely to shape expressions of specific emotions in academic settings, it is important to understand the dimensions of various achievement emotions in such contexts and how emotions relate to theorized antecedents and consequences. To date, most cross-cultural studies of achievement emotions have focused on the differences in the frequency and intensity of various emotions; little is known about the dimension of achievement emotions beyond the theoretical taxonomy and how emotions relate to self-efficacy and engagement in Chinese secondary classrooms. Understanding how achievement emotions manifest and function in a Chinese context can advance motivational theory by attending to emotion and motivation in different cultural contexts (Zusho & Kumar, 2018).

Current Study

The current study examines achievement emotions in Chinese high school chemistry classrooms and examines their relations with chemistry self-efficacy and engagement. In this study, the data analysis guided by two research questions: (1) What are the dimensions of Chinese students' achievement emotions in high school chemistry classrooms? (2) Do achievement emotions mediate

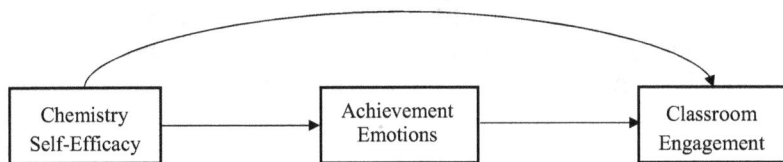

```
┌──────────────┐        ┌──────────────┐        ┌──────────────┐
│  Chemistry   │───────▶│ Achievement  │───────▶│  Classroom   │
│ Self-Efficacy│        │   Emotions   │        │  Engagement  │
└──────────────┘        └──────────────┘        └──────────────┘
```

Figure 2.1 The proposed theoretical model.

the relationship between chemistry self-efficacy beliefs and classroom engagement? Regarding the first question, I hypothesized that students' achievement emotions would have different dimensions and these dimensions may differ from existing findings in Western cultures. Given the scarcity of research on achievement emotions in Chinese context, however, I do not hypothesize about the specific nature of the dimensions. Regarding the relation of achievement emotions with chemistry self-efficacy and classroom engagement, I tested a model (shown in Figure 2.1) in which achievement emotions would partially mediate the relationship between chemistry self-efficacy and classroom engagement. Guided by control-value theory (Pekrun, 2006) and empirical findings (Pekrun et al., 2004, 2011; Walker et al., 2006), I expected (1) students' chemistry self-efficacy beliefs and achievement emotions would be directly related to classroom engagement; (2) chemistry self-efficacy would predict achievement emotions; and (3) different types of achievement emotions would be associated with classroom engagement in different ways.

Method

Participants

The study was conducted in a high school in a northern province in China. Participants were 103 11th grade students (45 female and 58 male) from two advanced classes as part of a larger project investigating the integration of computer simulations into science instruction. Students were 17 or 18 years old with Han ethnicity and their parents engaged in various professions such as government officers, teachers, and self-employed small-business owners. The two classes were equivalent in the average performance based on standardized school assessments. A multivariate analysis of variance (MANOVA) revealed no significant differences between the two classes with respect to four variables in the path model (F [4, 84] = 1.901, Wilk's Λ = .917; p = .118), and therefore I combined participants from the two classes into a single sample.

Instructional Context

To describe the instructional context for the study, the first author observed the chemistry class for two sessions. In each highly structured classroom, students sat row by row with limited free spaces. There were four chemistry classes in each week and each class period lasted 45 minutes. Students stayed in one classroom to take different courses while teachers traveled between different classrooms. During the typical chemistry instruction, chemistry teachers who used the same centralized curriculum standards, textbooks, and teaching materials usually stood on the platform and conducted lecture-dominated instruction that was supplemented with demonstration experiments and videos. Teachers usually interacted with students in the format of choral responding. The teacher-directed instruction and classroom structure suggested that the teacher-student relationship was hierarchical, and students were expected to follow teachers' directions.

Procedures and Measures

This study was conducted in the spring semester of 2016–2017 academic year. Participants completed a pencil-and-paper questionnaire that consisted of measures of chemistry self-efficacy, achievement emotions, and classroom engagement. Measures were group administered by chemistry teachers to each class at the beginning of one chemistry class in the same week. All students from the two advanced classes filled out the questionnaire without missing data. Responses were indicated using a 5-point Likert scale, anchored at 1 (strongly disagree) and 5 (strongly agree).

Achievement Emotions. Achievement emotions were assessed with the Chinese version of the Achievement Emotions Questionnaire – Mathematics (AEQ-M; Pekrun et al., 2011; Pekrun, Goetz, & Frenzel, 2005). Pekrun and colleagues translated the AEQ-M from German to Chinese and developed evidence supporting its validity for measuring achievement emotions in Chinese contexts. The AEQ-M includes subscales that assess achievement emotions in different contexts (e.g., doing homework, taking tests); given our research aims, I used the subscale that addresses emotions while attending class. I modified survey items for the chemistry classroom setting by changing "math" to "chemistry". (Appendix Part 1). The final scale consisted of 14 items (α = .794). An example item was: "I look forward to my chemistry classes".

Chemistry self-efficacy. Chemistry self-efficacy was assessed with a translation of Dalgety, Coll, and Jones's (2003) chemistry self-efficacy scale (Appendix Part 2). The first author translated the scale to Chinese and then one chemistry teacher who was fluent in English back-translated the Chinese version to English, and the first author verified the match with the English version. The scale consisted of 16 items (α = .942). An example item was: "I can convert the data obtained in a chemistry experiment into a result". Mean scores were used in analyses.

Classroom engagement. Classroom engagement was assessed with Skinner et al.'s (2008) measure of students' engagement in chemistry classes (Appendix Part 3). I used the same language translation process described for the self-efficacy measure. The scale consisted of 10 items (α = .896). One example item was: "When I'm in chemistry class, I think about other things [reverse coded]".

Analytic Plan

To identify dimensions of achievement emotions in the sample, I conducted exploratory factor analysis with SPSS version 22. Since different achievement emotions are assumed to be correlated, I used principal axis factoring with a direct oblimin rotation. Scree plots, Eigen values, and pattern matrix were examined to determine the number of latent factors. Prior to the extraction of the factors, Kaiser Meyer Olkin (.777) and Bartlett's Test of sphericity ($\chi^2 = 477.78$, $df = 91$, $p < .001$) indicated respondent data were suitable for factor analysis.

I then conducted path analysis with Mplus version 7.31 to assess the hypothesized relations between chemistry self-efficacy, achievement emotions, and classroom engagement with maximum likelihood estimation. I used absolute, parsimonious, and incremental indices to evaluate the data-model fit, including the model χ^2 statistic, the standardized root measure squared residual (SRMR), the root mean squared error of approximation (RMSEA), the Tucker-Lewis index (TLI), and the comparative fit index (CFI). I adopted Hu and Bentler's (1999) suggested values for retaining a model: RMSEA below .06, SRMR below .08, TLI above .95, and CFI above .95.

Results

Exploratory Factor Analysis (EFA)

With regard to the dimensions of Chinese students' achievement emotions, EFA results suggested a two-factor solution, which explained 40.26% of the variance. The descriptive statistics of the 14 items used in the analysis and the factor loading matrix for the final solution are presented in Table 2.2. One item (item 9), which had weak loadings (< .4) on both factors, was excluded from analysis. All remaining items had moderate to strong loadings on a single factor and no cross loadings greater than .4.

Factor 1, which I labeled *positive emotions*, was characterized by positive loadings for items measuring feelings of enjoyment (E1, E2, E3, E4) and pride (E5, E14), and negative loadings for items measuring feelings of anger (E6, E7, E8) and anxiety (E10). According to the taxonomy in Table 2.1, Factor 1 included two types of positive activating emotions and two types of negative activating emotions. Factor 2, which I labeled *shame*, one specific type of negative activating emotion, was characterized by three items about feeling shame (E11, E12, E13). In sum, I concluded that there were two distinct factors in Achievement Emotions measure with this sample of Chinese high school students in chemistry classrooms: *positive emotions* and *shame*.

Model Testing

Descriptive statistics and bivariate correlations between model variables are presented in Table 2.3. Chinese students had moderate levels of self-efficacy,

Table 2.2 Descriptive statistics and factor loading matrix of items in the achievement emotion scale

Achievement Emotion Scale Items	M	SD	Factor	
			1	*2*
E2 – I enjoy my chemistry classes.	3.60	.73	.833	
E1 – I look forward to my chemistry classes.	3.47	.78	.728	
E3 – The material we deal with in chemistry is so exciting that I really enjoy my classes.	3.32	.82	.706	
E6 – I am annoyed during my chemistry classes.	2.20	.83	−.683	
E10 – When thinking about my chemistry class, I get nervous.	1.98	.74	−.630	
E4 – I enjoy my class so much that I am strongly motivated to participate.	3.44	.74	.589	
E14 – I think I can be proud of my knowledge in chemistry.	3.82	.76	.587	
E5 – I am proud of my contributions to the chemistry class.	3.70	.85	.578	
E7 – I am so angry during my chemistry class that I would like to leave.	1.62	.78	−.501	
E8 – I get angry because the material in chemistry is so difficult.	2.38	.96	−.431	
E12 – My face is getting hot because I am embarrassed that I cannot answer the teacher's questions.	2.43	1.01		.754
E11 – When I say something in my chemistry class, I can tell that my face gets red.	2.23	.95		.627
E13 – I am ashamed that I cannot answer my chemistry teacher's questions well.	3.16	1.12		.561
E9 – I worry if the material is much too difficult for me.	3.14	1.05		

Note: Factor loading < .4 are not shown to highlight the factor structure.

Table 2.3 Descriptive statistics and bivariate correlations among four variables

	1 (PE)	*2 (S)*	*3 (SE)*	*4 (CE)*	Descriptives	
					M	SD
1. Positive Emotions (PE)	—				3.70	.55
2. Shame (S)	−.122	—			2.60	.80
3. Self-Efficacy (SE)	.390**	−.203*	—		3.32	.65
4. Classroom Engagement (CE)	.542**	−.144	.409**	—	3.67	.68

* Correlation is significant at the 0.05 level (2-tailed).
** Correlation is significant at the 0.01 level (2-tailed).

positive emotions, and engagement in chemistry classes. The perception of *shame* was slightly lower than *positive emotions*. According to Cohen's (1988) suggested criteria, bivariate correlations among *positive emotions*, self-efficacy, and engagement were positive and moderate while the correlation between *shame* and engagement was negative but not statistically significant.

Next, I tested the fit of an over-identified model in which *positive emotions* and *shame* partially mediate the relationship between chemistry self-efficacy and classroom engagement (Figure 2.2). Model fit indices demonstrated the model fitted the data well (χ^2 [1] = .975, p = .324, SRMR = .028, RMSEA < .001, CFI = 1.000, TLI = 1.003). The model accounted for significant amounts of variance in classroom engagement (34.7%) but non-significant amounts of variance in *positive emotions* (11.5%) and *shame* (8.9%). Figure 2.2 illustrates a model and shows significant path coefficients and residual variance of dependent variables. Table 2.4 presents a decomposition of all effects. Chemistry self-efficacy had a statistically significant, positive association with *positive emotions* (β = .338), and a

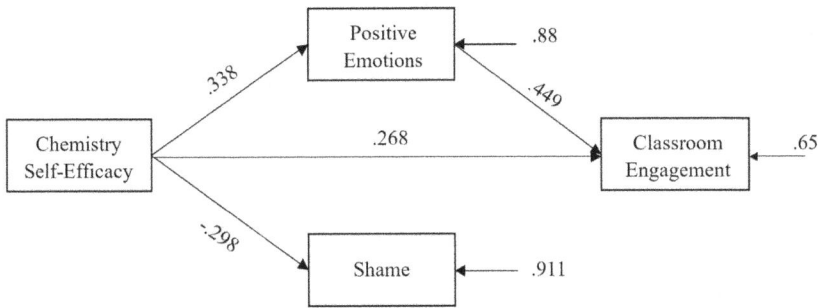

Figure 2.2 The path model.

Table 2.4 Maximum likelihood parameter estimates

Parameter	β	S.E.	Est./S.E.	p
Direct Effect				
Chemistry Self-Efficacy → Positive Emotions	.338	.098	3.437	.001
Chemistry Self-Efficacy → Shame	−.298	.105	−2.834	.005
Chemistry Self-Efficacy → Classroom Engagement	.268	.090	2.995	.003
Positive Emotions → Classroom Engagement	.449	.083	5.440	<. 001
Shame → Classroom Engagement	.038	.089	.430	.667
Indirect Effect				
Chemistry Self-Efficacy → Classroom Engagement	.152	.054	2.820	.005
Residual Variance				
Classroom Engagement	.653	.079	8.284	<. 001
Positive Emotions	.885	.067	13.286	<. 001
Shame	.911	.063	14.579	<. 001

statistically significant, negative association with *shame* ($\beta = -.298$). Chemistry self-efficacy had statistically significant, direct positive association ($\beta = .268$) with classroom engagement. Chemistry self-efficacy also had statistically significant indirect effect on classroom engagement via *positive emotions* ($\beta = .152$). *Positive emotions* had a statistically significant association with classroom engagement ($\beta = .449$), while *shame* was not significantly associated with engagement after controlling for *positive emotions*. Based on Suhr's (2008) criteria, all direct path coefficients ranged from medium to large.

Discussion

This study examined the dimensions of students' achievement emotions in highly structured Chinese chemistry classrooms. I identified two latent factors – *positive emotions* and *shame* – and examined the extent to which these factors mediated the relationship between self-efficacy and classroom engagement. I found that *positive emotions* partially mediated the effect of chemistry self-efficacy on engagement. Chemistry self-efficacy was also negatively associated with *shame*. In the following section, I discuss these findings in light of the role that cultural factors play in influencing Chinese students' emotional perceptions and their relationships with other variables in control-value theory. I also suggest implications for teachers regarding how to utilize emotional resources to effectively engage students in Eastern chemistry classrooms.

Dimensions of Achievement Emotions

A key finding of the study is that Chinese students' achievement emotions in chemistry classes had two distinct factors: *positive emotions* and *shame*. The first factor – *positive emotions* – refers to high levels of enjoyment and pride and low levels of anger and anxiety. Students' four class-related achievement emotions (i.e., enjoyment, pride, anger, and anxiety) were not discrete and isolated, but rather tended to highly correlate with each other. In contrast, items tapping a specific negative emotion – *shame* –constituted a unique pattern of correlations. This pattern suggests that Chinese high school students' perceptions of positive emotions were linked with perceptions of low levels of anger and anxiety, consistent with the notion that positive emotions (i.e., enjoyment and pride) and specific negative emotions (i.e., anger and anxiety) might constitute opposite ends of a bipolar spectrum (Pekrun et al., 2011). Students might have associated low levels of anger or anxiety with positive emotions because low-activating emotions (e.g., calmness) are valued while specific negative high-activating emotions (e.g., anger) are discouraged in Eastern classroom contexts (Heine, 2001; Wong & Tsai, 2007).

In our sample, *shame* emerged as a separate emotional factor, and one that was distinct from other negative emotions. This pattern is distinct from prior research in Western contexts that has grouped shame with other negative emotions (Pekrun et al., 2002a; Pekrun, 2006; Putwain et al., 2013; Shaver, Wu, & Schwartz, 1992). Why does *shame* stand out as an independent factor of Chinese students' achievement emotions in chemistry classes? Shame is a social emotion linked to students' beliefs about failure to meet others' expectations as part of the classroom community. The salience of shame as a unique negative emotion among Chinese students may reflect Eastern values that construct a student's sense of self. Confucian cultures, which stress the value of academic accomplishments and attribute personal responsibility for failures, create initial conditions for perceptions of shame (Turner & Waugh, 2007). Simultaneously, collectivistic cultures, which highlight obligation to others and the harmony of the whole group, expect students to carefully control strong negative emotions (e.g., anger) that might influence their relations with others negatively. Negative emotions such as anger, which are accepted in individualistic cultures, are more likely to be avoided in collectivistic culture (Frenzel et al., 2007). In contrast, shame is perceived as a more appropriate response than anger in Eastern contexts (Wong & Tsai, 2007). As a consequence, students in collectivistic cultures may be more likely to express shame, which is strongly tied to fulfilling the expectations of others (e.g., not being able to answer a teacher's questions), rather than anger, which emphasizes distance from others (Kitayama, Markus, & Kurokawa, 2000). The prominence of feelings of shame among Asian students (Mesquita, Boiger, & De Leersnyder, 2017) may be related to its corresponding social functions or demands in Eastern contexts.

Despite the fact that the feeling of *shame* is negative in the valence, such perceptions are also activating, which might lead to positive educational outcomes. Chinese parents or teachers use educational techniques to trigger the feeling of *shame* in students than Americans (Fung, Lieber, & Leung, 2003). In some Chinese classrooms, a misbehaved student may be asked to sit up front near the teacher's desk to help him or her regulate classroom behaviors. Chinese teachers often ask an inattentive student to stand up to answer questions. Such public attention from teachers and classmates can incite feelings of shame due to the failure of meeting classroom discipline. In summary, feeling *shame*, which is viewed as a valued and appropriate response to failures in Eastern cultures (Wong & Tsai, 2007), is more commonly experienced, salient, or threatening for Asian students (Heine, 2001).

Relations among Self-Efficacy, Achievement Emotions, and Engagement

Another important finding of the study was the relations among students' judgments about their capabilities to learn chemistry, achievement emotions, and engagement. Students with higher chemistry self-efficacy beliefs reported higher *positive emotions* and lower feelings of *shame*, and greater engagement in chemistry classes, consistent with findings of previous research conducted in

Western settings (Pekrun et al., 2004; Turner & Schallert, 2001). This consistency suggests these relations may be universal across cultural and subject contexts, which reflects Pekrun's (2006) argument that causal mechanisms of achievement emotions follow general principles.

Our findings of distinct patterns in how *positive emotions* and *shame* related to engagement add to mixed findings in the literature. Consistent with control-value theory and prior research in Western contexts (Pekrun et al., 2002a, 2011), students in our sample with a higher level of positive emotions reported more engagement in chemistry classes. However, shame was not significantly associated with classroom engagement, either in bivariate correlations or when controlling for positive emotions. For example, Pekrun et al. (2004, 2011) found that *shame* was significantly and negatively correlated with student engagement. Nevertheless, our findings are consistent with Lam, Wong, Yang, and Liu (2012), who found a low correlation ($r = -.03$) between negative emotions and student engagement in a large sample of Chinese middle school students. It should be noted that our findings do not indicate that shame was not associated with classroom engagement. As shame is negative but activating in nature and is more salient to Chinese students, the perception of shame may serve as a double-edged sword for students to regulate their following behaviors or strategies. Shame may prompt some students to spend more effort and process information more carefully in order to avoid situations of losing face and maintain the self-value for others. However, shame might also be a real emotional threat, which results in task-irrelevant thinking (Pekrun et al., 2004) and decreases engagement in classrooms. I suspect the relationship between negative activating emotions such as shame and classroom engagement may be complex and sensitive to cultural norms and individual differences, and thus represents a productive area for future research.

Overall, our findings underscore the importance of self-efficacy and positive emotions in supporting student engagement. This finding is consistent with prior research indicating the unique contribution of self-efficacy to classroom engagement when controlling for other variables (Walker et al., 2006). It also suggests that teachers can more effectively engage students in chemistry classrooms if pedagogical strategies can promote students' self-efficacy beliefs and positive emotions. For example, if teachers can provide scaffolding and assign a task at the appropriate difficulty level, students may feel more efficacious and engaged in classrooms (Lam et al., 2012).

Limitations and Conclusions

The finding of the current study should be considered in light of the following limitations. First, causal relationships among self-efficacy, achievement emotions, and classroom engagement cannot be established because the survey data was collected at a single time point. I suggest that future research collect similar data from a larger sample across multiple time points and explore causal relationships between different affective variables. Second, the finding of this study may reflect particular features of the sample (e.g., advanced students, chemistry context), and therefore I encourage research that attempts to replicate findings with other Chinese samples. In addition, not all achievement emotions were assessed (e.g., boredom, deactivating emotions) and only in-class emotions were examined. Future research that examines a wider range of emotions in different cultural contexts holds promising for elaborating understanding of how different achievement emotions function in such settings.

In conclusion, this study examined the dimensions, antecedent, and effect of achievement emotions in Chinese high school chemistry classrooms and its findings underscore the linkages between achievement emotions, self-efficacy, and classroom engagement. Our findings make two key contributions to psychological literature on achievement emotions. First, our study contributes to theoretical understandings of Chinese students' patterns of achievement emotions experienced in high school chemistry classrooms. Our findings indicated that *shame* was perceived differently than other types of negative emotions and argued the uniqueness of shame reflected cultural norms in Chinese classrooms that emphasize social relations with others and value the function of shame for motivating students to spend more effort. Second, I tested theoretical assumptions of control-value theory in a cultural and subject context that has received little attention. I found that *positive emotions*, but not *shame*, mediated the relationship between self-efficacy and classroom engagement. The non-significant association between *shame* and classroom engagement highlights an area that merits future attention to understand how negative emotions relate to student engagement, and how cultural or classroom contexts influence such relations. Our study adds to understanding of student emotions across cultural contexts with different values and educational systems. Understanding functional linkages of achievement emotions in highly structured Chinese classrooms may inform future research to investigate how teachers can develop effective pedagogical strategies (e.g., using technology) that nurture positive emotions and engage students in such environments.

Acknowledgements

The original version of this chapter was published in *International Journal of Science Education*. The full article can be found here: https://www.tandfonline.com/doi/abs/10.1080/09500693.2020.1734684

Appendix: Three Surveys Measuring Achievement Emotions, Self-Efficacy Beliefs, and Classroom Engagement

Part 1. This part pertains to the feelings you may experience when attending chemistry classes. Please carefully read each statement and decide to what extent it is true for you.

	SD	D	N	A	SA
1. I look forward to my chemistry classes.	O	O	O	O	O
2. I enjoy my chemistry classes.	O	O	O	O	O
3. The material we deal with in chemistry is so exciting that I really enjoy my classes.	O	O	O	O	O
4. I enjoy my class so much that I am strongly motivated to participate.	O	O	O	O	O
5. I am proud of my contributions to the chemistry class.	O	O	O	O	O
6. I am annoyed during my chemistry classes.	O	O	O	O	O
7. I am so angry during my chemistry class that I would like to leave.	O	O	O	O	O
8. I get angry because the material in chemistry is so difficult.	O	O	O	O	O
9. I worry if the material is much too difficult for me.	O	O	O	O	O
10. When thinking about my chemistry class, I get nervous.	O	O	O	O	O
11. When I say something in my chemistry class, I can tell that my face gets red.	O	O	O	O	O
12. My face is getting hot because I am embarrassed that I cannot answer the teacher's questions.	O	O	O	O	O
13. I am ashamed that I cannot answer my chemistry teacher's questions well.	O	O	O	O	O
14. I think I can be proud of my knowledge in chemistry.	O	O	O	O	O

Notes:
SD – Strongly Disagree
D – Disagree
N – Neutral
A – Agree
SA – Strongly Agree

Part 2. This part investigates the confidence you have in undertaking different tasks. Please carefully read each statement and decide to what extent it is true for you.

	SD	D	N	A	SA
1. I can apply a set of chemistry rules to different elements of the Periodic Table.	O	O	O	O	O
2. I can achieve a passing grade in a chemistry test.	O	O	O	O	O
3. I can tutor another student in class.	O	O	O	O	O
4. I can ensure that data obtained from an experiment is accurate.	O	O	O	O	O
5. I can propose a meaningful question that could be answered experimentally.	O	O	O	O	O

	SD	D	N	A	SA
6. I can explain something that I learned in this chemistry course to another person.	○	○	○	○	○
7. I can choose an appropriate formula to solve a chemistry problem.	○	○	○	○	○
8. I know how to convert the data obtained in a chemistry experiment into a result.	○	○	○	○	○
9. After reading an article about a chemistry experiment, I can write a summary of the main points.	○	○	○	○	○
10. I can learn and explain chemistry theory.	○	○	○	○	○
11. I can determine the appropriate units for a result determined using a formula.	○	○	○	○	○
12. I can write up the experimental procedures in a laboratory report	○	○	○	○	○
13. After watching a television documentary dealing with some aspect of chemistry, I can write a summary of its main points.	○	○	○	○	○
14. I can apply theory learned in a lecture for a laboratory experiment.	○	○	○	○	○
15. I can write up the results section in a laboratory report.	○	○	○	○	○
16. After listening to a public lecture regarding some chemistry topic, I can explain its main ideas to another person.	○	○	○	○	○

Part 3. This part pertains to engagement when attending chemistry classes. Please carefully read each statement and decide to what extent it is true for you.

	SD	D	N	A	SA
1. I try hard to do well in chemistry class.	○	○	○	○	○
2. In chemistry class, I work as hard as I can.	○	○	○	○	○
3. When I'm in chemistry class, I participate in class discussions.	○	○	○	○	○
4. I pay attention in chemistry class.	○	○	○	○	○
5. When I'm in chemistry class, I listen very carefully.	○	○	○	○	○
6. When I'm in chemistry class, I just act like I'm working.	○	○	○	○	○
7. I don't try very hard in chemistry class.	○	○	○	○	○
8. In chemistry class, I do just enough to get by.	○	○	○	○	○
9. When I'm in chemistry class, I think about other things.	○	○	○	○	○
10. When I'm in chemistry class, my mind wanders.	○	○	○	○	○

References

Bandura, A. (1986). The explanatory and predictive scope of self-efficacy theory. *Journal of Social and Clinical Psychology*, *4*, 359–373. https://doi.org/10.1521/jscp. 1986.4.3.359

Bandura, A. (1997). *Self-efficacy: The exercise of control*. New York: Freeman.

Bandura, A. (2006). Guide for constructing self-efficacy scales. In F. Pajares & T. Urdan (Eds.), *Self-efficacy beliefs of adolescents* (pp. 307–337). Greenwich, CT: Information Age.

Bond, M. H., Leung, K., & Wan, K. C. (1982). The social impact of self-effacing attributions: The Chinese case. *The Journal of Social Psychology*, *118*, 157–166. https://doi.org/10.1080/00224545.1982.9922794

Cohen, J. (1988). *Statistical power analysis for the behavioral sciences* (2nd ed.). Hillsdale, NJ: Lawrence Erlbaum.

Dalgety, J., Coll, R. K., & Jones, A. (2003). Development of Chemistry Attitudes and Experiences Questionnaire (CAEQ). *Journal of Research in Science Teaching*, *40*, 649–668. https://doi.org/10.1002/tea.10103

Ding, M., Li, Y., Li, X., & Kulm, G. (2008). Chinese teachers' perceptions of students' classroom misbehaviour. *Educational Psychology*, *28*, 305–324. https://doi.org/10.1080/01443410701537866

Eid, M., & Diener, E. (2001). Norms for experiencing emotions in different cultures: Inter-and intranational differences. *Journal of Personality and Social Psychology*, *81*, 869. https://doi.org/10.1037//0022-3514.81.5.869

Frenzel, A. C., Thrash, T. M., Pekrun, R., & Goetz, T. (2007). Achievement emotions in Germany and China: A cross-cultural validation of the Academic Emotions Questionnaire—Mathematics. *Journal of Cross-Cultural Psychology*, *38*, 302–309. https://doi.org/10.1177/0022022107300276

Fung, H., Lieber, E., & Leung, P. W. (2003). Parental beliefs about shame and moral socialization in Taiwan, Hong Kong and the United States. In K. S. Yang et al. (Eds.), *Progress in Asian social psychology: Conceptual and empirical contributions* (pp. 83–109). Westport, CT: Praeger.

Gow, L., Balla, J., Kember, D., & Hau, K. T. (1996). The learning approaches of Chinese people: A function of socialization processes and the context of learning. In M. H. Bond (Ed.), *The handbook of Chinese psychology* (pp. 109–123). Oxford, England: Oxford University Press.

Heine, S. J. (2001). Self as cultural product: An examination of East Asian and North American selves. *Journal of Personality*, *69*, 881–905. https://doi.org/10.1111/1467-6494.696168

Hsu, F. L. (1971). Psychosocial homeostasis and jen: Conceptual tools for advancing psychological anthropology. *American Anthropologist*, *73*, 23–44. https://doi.org/10.1525/aa.1971.73.1.02a00030

Hu, H. C. (1944). The Chinese concepts of "face". *American Anthropologist*, *46*, 45–64. https://doi.org/10.1525/aa.1944.46.1.02a00040

Hu, L. T., & Bentler, P. M. (1999). Cutoff criteria for fit indexes in covariance structure analysis: Conventional criteria versus new alternatives. *Structural Equation Modeling: A Multidisciplinary Journal*, *6*, 1–55. https://doi.org/10.1080/10705519909540118

Kitayama, S., Markus, H. R., & Kurokawa, M. (2000). Culture, emotion, and well-being: Good feelings in Japan and the United States. *Cognition & Emotion*, *14*, 93–124. https://doi.org/10.1080/026999300379003

Klassen, R. M. (2004). Optimism and realism: A review of self-efficacy from a cross-cultural perspective. *International Journal of Psychology*, *39*, 205–230. https://doi.org/10.1080/00207590344000330

Lam, S. F., Wong, B. P., Yang, H., & Liu, Y. (2012). Understanding student engagement with a contextual model. In S. L. Christenson, A. L. Reaschly, & C. Wylie (Eds.), *Handbook of research on student engagement* (pp. 403–419). New York, NY: Springer. http://dx.doi.org/10.1007/978-1-4614-2018-7_19

Lee, J. (2009). Universals and specifics of math self-concept, math self-efficacy, and math anxiety across 41 PISA 2003 participating countries. *Learning and Individual Differences*, *19*, 355–365. https://doi.org/10.1016/j.lindif.2008.10.009

Li, J., Wang, L., & Fischer, K. (2004). The organisation of Chinese shame concepts?. *Cognition and Emotion*, *18*, 767–797. https://doi.org/10.1080/02699930341000202

Linnenbrink, E. A., & Pintrich, P. R. (2003). The role of self-efficacy beliefs in student engagement and learning in the classroom. *Reading & Writing Quarterly*, *19*, 119–137. https://doi.org/10.1080/10573560308223

Luo, W., Ng, P. T., Lee, K., & Aye, K. M. (2016). Self-efficacy, value, and achievement emotions as mediators between parenting practice and homework behavior: A control-value theory perspective. *Learning and Individual Differences*, *50*, 275–282. https://doi.org/10.1016/j.lindif.2016.07.017

Marchand, G. C., & Gutierrez, A. P. (2012). The role of emotion in the learning process: Comparisons between online and face-to-face learning settings. *The Internet and Higher Education*, *15*, 150–160. https://doi.org/10.1016/j.iheduc.2011.10.001

Markus, H. R., & Kitayama, S. (1991). Culture and the self: Implications for cognition, emotion, and motivation. *Psychological Review*, *98*, 224. https://doi.org/10.1037//0033-295x.98.2.224

Marsella, A. J., Murray, M. D., & Golden, C. (1974). Ethnic variations in the phenomenology of motions: I. Shame. *Journal of Cross-Cultural Psychology*, *5*, 312–328. https://doi.org/10.1177/002202217400500305

Mesquita, B., Boiger, M., & De Leersnyder, J. (2017). Doing emotions: The role of culture in everyday emotions. *European Review of Social Psychology*, *28*, 95–133. https://doi.org/10.1080/10463283.2017.1329107

Mesquita, B., & Karasawa, M. (2002). Different emotional lives. *Cognition & Emotion*, *16*, 127–141. https://doi.org/10.1080/0269993014000176

Oyserman, D., Coon, H. M., & Kemmelmeier, M. (2002). Rethinking individualism and collectivism: Evaluation of theoretical assumptions and meta-analyses. *Psychological Bulletin*, *128*, 3–72. https://doi.org/10.1037/0033-2909.128.1.3

Pajares, F., & Miller, M. D. (1994). Role of self-efficacy and self-concept beliefs in mathematical problem solving: A path analysis. *Journal of Educational Psychology*, *86*, 193. https://doi.org/10.1037//0022-0663.86.2.193

Pekrun, R. (2006). The control-value theory of achievement emotions: Assumptions, corollaries, and implications for educational research and practice. *Educational Psychology Review*, *18*, 315–341. https://doi.org/10.1007/s10648-006-9029-9

Pekrun, R., Goetz, T., & Frenzel, A. C. (2005). *Academic Emotions Questionnaire–Mathematics (AEQ-M): User's manual*. Munich, Germany: University of Munich, Department of Psychology.

Pekrun, R., Goetz, T., Frenzel, A. C., Barchfeld, P., & Perry, R. P. (2011). Measuring emotions in students' learning and performance: The Achievement Emotions Questionnaire (AEQ). *Contemporary Educational Psychology*, *36*, 36–48. https://doi.org/10.1016/j.cedpsych.2010.10.002

Pekrun, R., Goetz, T., Perry, R. P., Kramer, K., Hochstadt, M., & Molfenter, S. (2004). Beyond test anxiety: Development and validation of the Test Emotions Questionnaire (TEQ). *Anxiety, Stress & Coping, 17*, 287–316. https://doi.org/10.1080/10615 800412331303847

Pekrun, R., Goetz, T., Titz, W., & Perry, R. P. (2002a). Academic emotions in students' self-regulated learning and achievement: A program of qualitative and quantitative research. *Educational Psychologist, 37*, 91–105. https://doi.org/10.1207/s15326985 ep3702_4

Pekrun, R., Goetz, T., Titz, W., & Perry, R. P. (2002b). Positive emotions in education. In E. Frydenberg (Ed.), *Beyond coping: Meeting goals, visions, and challenges* (pp. 149–174). Oxford, UK: Elsevier.

Pekrun, R., & Linnenbrink-Garcia, L. (2012). Academic emotions and student engagement. In S. L. Christenson, A. L. Reschly, & C. Wylie (Eds.), *Handbook of research on student engagement* (pp. 259–282). New York, NY: Springer.

Putwain, D., Sander, P., & Larkin, D. (2013). Academic self-efficacy in study-related skills and behaviours: Relations with learning-related emotions and academic success. *British Journal of Educational Psychology, 83*, 633–650. https://doi.org/10.1111/ j.2044-8279.2012.02084.x

Putwain, D. W., & Symes, W. (2014). The perceived value of maths and academic self-efficacy in the appraisal of fear appeals used prior to a high-stakes test as threatening or challenging. *Social Psychology of Education, 17*, 229–248. https://doi.org/10.1007/ s11218-014-9249-7

Reschly, A. L., Huebner, E. S., Appleton, J. J., & Antaramian, S. (2010). Engagement as flourishing: The contribution of positive emotions and coping to adolescents' engagement at school and with learning. *Psychology in the Schools, 45*, 419–431. https://doi. org/10.1002/pits.20306

Russell, J. A., & Yik, M. S. M. (1996). Emotion among the Chinese. In M. H. Bond (Ed.), *The handbook of Chinese psychology* (pp. 166–188). Hong Kong: Oxford University Press.

Scholz, U., Doña, B. G., Sud, S., & Schwarzer, R. (2002). Is general self-efficacy a universal construct? Psychometric findings from 25 countries. *European Journal of Psychological Assessment, 18*, 242–251. https://doi.org/10.1027//1015-5759.18.3.242

Schunk, D. H. (1991). Self-efficacy and academic motivation. *Educational Psychologist, 26*(3–4), 207–231.

Schunk, D. H. (2001). Social-cognitive theory and self-regulated learning. In B. J. Zimmerman & D. H. Schunk (Eds.), *Self-regulated learning and academic achievement: Theoretical perspectives* (2nd ed., pp. 125–151). Mahwah, NJ: Lawrence Erlbaum Associates, Inc.

Schutz, P. A., & Pekrun, R. (Eds.). (2007). *Emotion in education.* San Diego, CA: Academic Press.

Shaver, P. R., Wu, S., & Schwartz, J. C. (1992). Cross-cultural similarities and differences in emotion and its representation: A prototype approach. In M. S. Clark (Ed.), *Review of personality and social psychology* (Vol. 13, pp. 175–212). Newbury Park, CA: Sage.

Skinner, E., Furrer, C., Marchand, G., & Kindermann, T. (2008). Engagement and disaffection in the classroom: Part of a larger motivational dynamic? *Journal of Educational Psychology, 100*, 765–781. https://doi.org/10.1037/a0012840

Stigler, J. W., & Perry, M. (1988). Mathematics learning in Japanese, Chinese, and American classrooms. In G. B. Saxe & M. Greathatt (Eds.), *Children's mathematics: New directions for child development*, No. 41 (pp. 27–53). San Francisco: Jossey-Bass.

Suhr, D. (2008). *Step your way through path analysis. Annual Conference Proceedings, Western Users of SAS Software*. CA: Universal City.

Tsai, J. L., Knutson, B., & Fung, H. H. (2006). Cultural variation in affect valuation. *Journal of Personality and Social Psychology*, *90*, 288–307. https://doi.org/10.1037/e633912013-121

Turner, J. E., & Schallert, D. L. (2001). Expectancy-value relationships of shame reactions and shame resiliency. *Journal of Educational Psychology*, *93*(2), 320–329. https://doi.org/10.1037/0022-0663.93.2.320

Turner, J. E., & Waugh, R. M. (2007). A dynamical systems perspective regarding students' learning processes: Shame reactions and emergent self-organizations. In P. Schutz & R. Pekrun (Eds.), *Emotions and education* (pp. 125–145). Burlington, MA: Academic.

Usher, E. L., & Pajares, F. (2008). Sources of self-efficacy in school: Critical review of the literature and future directions. *Review of Educational Research*, *78*, 751–796. https://doi.org/10.3102/0034654308321456

Walker, C. O., Greene, B. A., & Mansell, R. A. (2006). Identification with academics, intrinsic/extrinsic motivation, and self-efficacy as predictors of cognitive engagement. *Learning and Individual Differences*, *16*, 1–12. https://doi.org/10.1016/j.lindif.2005.06.004

Wigfield, A., & Cambria, J. (2010). Students' achievement values, goal orientations, and interest: Definitions, development, and relations to achievement outcomes. *Developmental Review*, *30*, 1–35. https://doi.org/10.1016/j.dr.2009.12.001

Wilson, R. W. (1981). Moral behavior in Chinese society: A theoretical perspective. In R. Wilson, S. Greenblatt, & A. Wilson (Eds.), *Moral behavior in Chinese society* (pp. 117–136). New York: Praeger.

Wong, Y., & Tsai, J. (2007). Cultural models of shame and guilt. In J. L. Tracy, R. W. Robins, & J. P. Tangney (Eds.), *The self-conscious emotions: Theory and research* (pp. 209–223). New York, NY: Guilford Press.

Zembylas, M. (2003). Emotions and teacher identity: A poststructural perspective. *Teachers and Teaching*, *9*, 213–238. https://doi.org/10.1080/13540600309378

Zimmerman, B. J. (2000). Self-efficacy: An essential motive to learn. *Contemporary Educational Psychology*, *25*, 82–91. https://doi.org/10.1006/ceps.1999.1016

Zusho, A., & Kumar, R. (2018). Introduction to the special issue: Critical reflections and future directions in the study of race, ethnicity, and motivation. *Educational Psychologist*, *53*, 61–63. https://doi.org/10.1080/00461520.2018.1432362

Zusho, A., & Pintrich, P. R. (2003). A process-oriented approach to culture: Theoretical and methodological issues in the study of culture and motivation. In F. Salili & R. Hoosain (Eds.), *Teaching, learning, and motivation in a multicultural context* (pp. 33–65). Greenwich, CT: Information Age.

3 Assessing Students' Achievement Emotions during Chemistry Experiments

Background

Doing experiments is one important pedagogical strategy in science education to help students connect macroscopic phenomena with microscopic mechanisms, test scientific hypotheses with observed evidence, develop scientific inquiry skills, and finally achieve conceptual understanding of scientific knowledge and the nature of science (Hofstein & Lunetta, 2003). Its distinctive role is also highlighted in Compulsory Science Curriculum Standards in China (The Ministry of Education, 2022). Doing experiments is thought to be a scientific method that enables students to control variables, collect information, and justify assertions. In the chemistry domain, laboratory activities also allow teachers to remind students of appropriate ways of using chemicals, operating laboratory apparatus, and disposing of waste, which contribute to enhancing students' consciousness of safety and responsibilities of protecting the environment. At present, students' achievement emotions in this unique environment is underexamined. In the first two chapters, I reviewed previous literature and described how cultural values and subject domains shape students' perceptions of achievement emotions in academic settings. Compared with the regular classroom setting, the scenario of conducting chemistry experiments is content-based and activity-oriented. During this process, students can conduct rich hands-on activities with chemicals and experimental apparatus, observe scientific phenomena, collaborate with group members, and enjoy the success or face the failure of experimental results. Just as described in control-value theory, all aforementioned elements related to activities and outcomes may incite students' various positive and negative emotions. Therefore, it is worth further investigating students' achievement emotions in this unique but complex context in chemistry teaching and learning.

Before investigating students' achievement emotions in doing chemistry experiments, it is necessary to describe the role or function of chemistry experiments in Eastern school contexts. In most Chinese high schools, students are expected to experience the procedure and learn basic operations of doing textbook experiments in chemistry labs. For example, each textbook unit across different grade levels are supplemented with one or two chemistry experiments,

DOI: 10.4324/9781032649665-5

which aim to help students understand or apply related theoretical knowledge in practices of doing experiments. All experimental activities listed in national chemistry textbooks are listed in Table 3.1.

As textbook experiments are relatively simple and there are no rigid requirements for the outcome of doing experiments (e.g., success or failure), lab environments are thought to be more relaxed and less stressful than regular

Table 3.1 Experiments listed in chemistry textbooks

School Level	Grade Level	Chemistry Experiments
Middle school	9th grade	• The production of oxygen in chemistry labs and its properties; • The production of carbon dioxide in chemistry labs and its properties; • Conditions for the combustion; • The physical and chemical properties of metals; • The preparation of sodium chloride solution with specific mass fractions of solute; • The chemical properties of acid and alkali; • The examination of one solution's pH value; • The removal of insoluble impurities from crude salt;
High School	10th grade	• The preparation of solutions with specific molarities; • The properties of iron and its compound; • The dynamic change of properties of elements in the same group on the periodic table; • The removal of impurities from crude salt with the precipitation method; • The transformation of sulfur compounds with different valences; • The transformation from chemical energy to electric energy; • The influencing factors of chemical reaction rate; • The identification of structures of organic compounds with ball-and stick models; • The properties of ethanol and acetic acid;
	11th grade	• The factors influencing the shift of chemical equilibrium; • The neutralization titration between strong acid and strong base; • The application of the hydrolysis of salts; • The simple electroplating experiment; • The creation of simple fuel cells; • The formation of simple coordination compounds;
	12th grade	• The production of ethyl acetate and its properties; • The verification of typical functional groups in organic compounds; • The properties of saccharides.

classroom settings. There are two main reasons or rationales for examining students' achievement emotions in doing chemistry experiments. On one hand, as one essential component of chemistry education, the environment of doing chemistry experiments owns its unique characteristics, which potentially influence students' emotional experiences. For example, diverse experimental phenomena such as color changes, the massive bubble of gas, and precipitation may incite students' interests, curiosity, and positive feelings toward natural worlds and the subject of chemistry. Simultaneously, safety concerns including the interaction with toxic chemicals and potential explosions from violent disruptive reactions may lead to feelings of fear and nervousness. On the other hand, the collaboration and competition existing among group members in experimental environments provide rich opportunities for inciting and influencing students' positive and negative emotions. Minkley, Ringeisen, Josek, and Kärner (2017) investigated students' various emotions experienced in three experiment-related conditions: Watching a biology experiment on video (i.e., the passive condition), conducting experiments on their own (i.e., the active condition), or in small groups (i.e., the interactive condition). Results showed that students in the active condition displayed the highest stress, followed by the interaction condition. In contrast, students in the passive condition experienced the least stress but with the cost of minimized enjoyment and maximized boredom. This randomized experimental field study has demonstrated that environmental characteristics such as the degree of subjective control and peer interaction may influence students' emotional experiences of doing experiments (Sinatra, Broughton, & Lombardi, 2014). At present, most of existing research has focused on the individual's activity-related or outcome-related emotions in an isolated setting and ignores emotions derived from the relevant social context surrounding him or her (Hareli & Weiner, 2002). To address this issue, Graham and Taylor (2014) used the attribution theory as a lens to distinguish between various emotions: Some emotions are related to feelings about oneself (e.g., pride and shame) while other emotions are related to feelings about other people (e.g., anger). In fact, social interactions with peers and teachers may have direct influence on students' perceived emotions. For example, if the group collaboration in doing chemistry experiments is very smooth, group members may feel the process is very efficient and perceive more positive emotions (e.g., happiness and gratitude) and less negative emotions (e.g., stress). In contrast, if one group member makes a big mistake in experimental operations, others may feel the mistake could be avoided and perceive negative emotions due to the failure of chemistry experiments (e.g., sadness and anger). What is more, the competition existing between different groups may also trigger various emotions. The group who lags behind peers may perceive anxiety while the leading group may perceive pride and happiness. These examples and scenarios demonstrate the diversity and complexity of emotions that students may experience in doing experiments.

However, currently, only a few studies have examined students' emotions in doing experiments. Randler, Hummel, and Wüst-Ackermann (2013) examined university students' situational disgust during dissection or when encountering living animals. Itzek-Greulich et al. (2017) reported that hands-on experience in chemistry lab environments enhanced the feeling of joy and reduced the feeling of boredom. Investigating students' achievement emotions in doing chemistry experiments has important implications for chemistry teaching and learning. Theoretically, it can help education researchers better understand how various emotions, specifically social-related emotions, are developed and expressed in such unique environments. Practically, the developed instrument can help chemistry teachers monitor students' learning experiences in chemistry laboratories and provide related pedagogical support for enhancing positive effects and reducing negative effects.

Item Development

Existing Questionnaires Measuring Achievement Emotions

The purpose of the present study was to develop a self-report measure of achievement emotions in conducting chemistry experiments with good validity and reliability. Following the construct of achievement emotions in control-value theory (Pekrun, 2006), I first reviewed the two well-recognized published questionnaires: The Chinese language version of the Achievement Emotion Questionnaire – Mathematics (AEQ-M) developed by Pekrun, Goetz, and Frenzel (2005) and Chinese Teenagers' Achievement Emotions Questionnaire (CTAEQ) developed by Dong and Yu (2007). Then all measured positive and negative emotions were listed, matched, and compared. The full description of definitions and dimensions of achievement emotions in control-value theory can be found in Chapter 2. Different words or terms used to describe students' feelings between the two surveys are italicized in Table 3.2. As shown in Table 3.2, main differences were located in categories of positive deactivating emotions (*relief* versus *calm*) and negative deactivating emotions (*disappointment* versus *tiredness-upset*). The possible explanation for these differences is that some emotions (e.g., calmness) are more valued and prevalent in

Table 3.2 Emotions measured in the two surveys

Valance	Degree of Activation	Pekrun (2006)	Dong and Yu (2007)
Positive	Activating	*Enjoyment*, Joy, Hope, Pride, *Gratitude*	Joy, Hope, Pride
	Deactivating	Relaxation, Contentment, *Relief*	Relaxation, Contentment, *Calm*
Negative	Activating	Anger, *Frustration*, Anxiety, Shame	Anger, Anxiety, Shame
	Deactivating	Boredom, Sadness, Hopelessness, *Disappointment*	Boredom, Sadness, Hopelessness, *Tiredness-upset*

Eastern cultural contexts (Mesquita & Karasawa, 2002). In order to identify students' possible emotions occurring in lab environments accurately, I incorporated all the aforementioned emotional words and tried to identify contextual elements that may trigger these positive or negative emotions.

Classroom Observations, Student Interviews, and Open-ended Responses

Before developing concrete survey items, classroom observations, semi-structured interviews, and open-ended responses were used to investigate students' experiences and feelings during the process of doing chemistry experiments and what contextual factors influenced their feelings. As Chinese students start to learn chemistry in ninth grade (i.e., the last year of middle school) in the K-12 education system, I first observed a ninth-grade chemistry class named "Exploring the candle burning". The class was selected as the object of observation because it was the first time that students had opportunities to do chemistry experiments. During the class, students were expected to observe and describe the phenomena and mechanism of candle burning in small groups.

During the observation process, I found that there were several episodes in which students clearly showed emotions or feelings in the formats of oral and body language, which constituted the whole classroom atmosphere. These moments were listed in the time sequence in Table 3.3. At the beginning of the class, the teacher first presented students' homemade videos of candle burning. Even though this scenario was a relatively passive setting where students simply watched others conducting experiments (Chi, 2009; Minkley et al., 2017), middle school students were excited and engaged because these videos were closely relevant to everyday life. Subsequently, the chemistry class evolved into an interactive setting where students worked in small groups to light the candle, observe phenomena and examine the product of combustion. Just as expected, students experienced mixed feelings including

Table 3.3 Class episodes where students displayed emotions

Episodes	Emotions
1. The chemistry teacher presented students' home-made videos of candle burning.	Students were *excited* and engaged because videos were self-made and derived from everyday life.
2. Students lighted the candle, observed phenomena, and examined the product of combustion in small groups.	Students were *curious* about chemical apparatus and were *happy* when they observed expected phenomena. Some students were *nervous* because they were not good at striking a match. They *worried* that incorrect operations may hurt group members.
3. Students conducted the experiment developed by British scientist Michael Faraday.	Students were *surprised* and *excited* when observing the second flame.

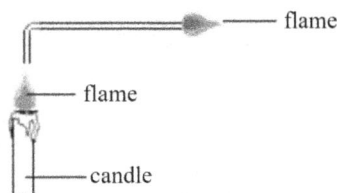

Figure 3.1 Faraday's candle experiment.

both positive emotions (e.g., happiness) and negative emotions (e.g., nervousness and worry), which derived from interactions with chemical apparatus and related safety concerns. At the end of the class, the teacher developed another interactive scenario where students were encouraged to explore Faraday's candle experiment (see Figure 3.1). They lighted the gas at the end of the glass tube and observed the emergence of the flame. As Faraday's candle experiment was not introduced in chemistry textbooks, students were very curious about possible results and showed much excitement when observing the second flame. The reexperience about how famous scientists conducted inquiry experiments also potentially incited students' enthusiasm for learning science. In summary, the classroom observation showed that experimental phenomena, operations of chemical apparatus, and social relationships within groups were three main sources that potentially shaped students' emotional experiences in chemistry lab classes. Positive emotions may include excitement, surprise, and happiness while negative emotions may include nervousness, fear, and anxiety.

Besides classroom observations, open-ended responses and semi-structured interviews were also used to examine students' experiences or feelings about doing chemistry experiments. Participants were recruited from ninth grade to 12th grade and 487 students responded to an open-ended question: What is your feeling about doing chemistry experiments? Several students were further interviewed to capture more detailed information. Examples of interview questions included: "Can you describe an impressive moment in chemistry labs? Have you ever worked with others to complete chemistry experiments? If yes, how do you feel about the group collaboration?" The collected qualitative data were organized and categorized based on common types of achievement emotions. As showed in Table 3.4, interview quotes and open responses implied that students perceived various positive emotions (e.g., enjoyment, satisfaction, relaxation) and negative emotions (e.g., anger, anxiety, fear) during the process of doing chemistry experiments simultaneously because lab learning environments were so distinct and complex. These mentioned emotions were basically consistent with those emotions identified in the classroom observation. Among them, some emotions were activity-related (e.g., relaxation, enjoyment, hectic, anger) while others were outcome-related (e.g., fear and satisfaction). Besides these commonly described emotions experienced in general classroom settings, some emotions such as fear and agitation are more

Table 3.4 Students' interview quotes and open-responses

Emotions	Quote
Enjoyment Joy Hope	Very interesting! Hope to do more experiments!
	I look forward to my lab classes. They are new and worth exploring.
	I like doing chemistry experiments because I can see different chemical reagents.
	Lab classes are interesting. Hope that we have more chances to take lab classes.
	I hope that chemical apparatus could be newer and more accurate and we could have more hands-on activities.
Satisfaction	I feel satisfied when experimental results fit with the expectation.
	Doing experiments can make me feel a sense of accomplishment.
Relaxation	Doing experiments is interesting and relaxing.
Anger	The calculation is so annoying. Sometimes [liquids] run outside [the container] and I doubt about my operational ability.
Anxiety	I need to look at textbooks and do experiments [simultaneously]. It is a *hectic* situation.
Fear	I am interested in doing experiments but some chemicals like sodium hydroxide are *frightening*.
	I like doing experiments and seeing beautiful phenomena but I am also scared of [potential] dangers.
Schadenfreude	I sometimes feel pleased when seeing others' failures in doing experiments.

salient in chemistry labs where students need to do hands-on activities with chemicals and apparatus.

In addition, one student said that he may perceive the feeling of envy if other groups had obvious phenomena but his group did not achieve such a goal. The emotion of envy was out of expectation, which promoted me to pay more attention to emotions derived from social interactions between the student self and others (e.g., peers and teachers). Existing research has described how two types of social-related emotion – envy and admiration – differentiate from each other. Envy is perceived when the student self desires advantages of others, particularly uncontrollable qualities (e.g., abilities) (Smith, Parrott, Diener, Hoyle, & Kim, 1999). Hareli and Weiner (2002) noted that envy was related to feelings of inferiority due to social comparisons and negative evaluations. In contrast, the other type of emotion – admiration – is perceived when the student self attributes others' success to high efforts. In Eastern cultural contexts, the boundary between admiration and envy is vague, which is reflected in a meme on the internet describing the dynamic change of human emotions: "admiration, envy, and hate". This meme shows that as opposite ends of a bipolar spectrum (Pekrun, Goetz, Frenzel, Barchfeld, & Perry, 2011), positive emotions may transform into negative emotions in some situations. The transformation depends on how individuals approach or attribute the success or failure of academic tasks.

Instrument Validation

The qualitative content analysis and existing survey items (e.g., AEQ and CTAEQ) provide rich resources for developing the new survey. Within the theoretical framework of control-value theory, we first identified 14 emotions associated with lab learning environments. The original scale which measured students' achievement emotions about doing chemistry experiments (AEQ-CE) consisted of 24 items with Cronbach $\alpha = 0.773$ (see Table 3.5), which described students' various feelings in doing experiments, data analysis, and

Table 3.5 Original survey items developed for measuring students' achievement emotions in chemistry experiments

Emotions	Survey Items
Enjoyment/Joy	Q1: I look forward to my lab classes.
	Q2: I feel excited when I can have chemistry classes in the lab.
	Q5: I feel glad when I can observe obvious phenomena.
	Q11: I feel happy when I can complete experiments quicker than my classmates.
	Q4: Phenomena observed in chemistry experiments are interesting.
	Q13: I feel happy when I can work with others to do chemistry experiments.
Calm	Q23: I can keep calm during chemistry lab exams.
Envy	Q12: I envy that others' experiments have obvious phenomena.
Relaxation	Q7: I feel relaxed when doing chemistry experiments.
Contentment	Q18: I feel satisfied when experiment results are in line with expectations.
Pride	Q20: I feel a sense of accomplishment when I can finish the experiment smoothly.
Anger	Q15: I feel angry when the experiment does not go well due to team members' mistakes.
	Q17: I feel annoyed when I have difficulty in analyzing the data.
Anxiety	Q6: I feel nervous when I do hands-on activities in the lab.
	Q9: I feel nervous when I am criticized by the chemistry teacher due to improper operations.
	Q22: I feel nervous before the experiment operation exam.
	Q19: I feel anxious when experimental results are out of expectation.
	Q10: I feel anxious when my experiment progress is slower than others.
Shame	Q14: I feel shamed when I hold back the progress of group experiments.
Fear	Q3: I feel afraid of taking lab classes because some chemicals are toxic.
	Q8: I may feel frightened if some accidents happened in the lab (e.g., breaking the glassware).
Boredom	Q16: I feel bored when waiting for the emergence of phenomena.
Sadness	Q21: I may feel sad if my experiment did not go well.
Schadenfreude	Q24: I sometimes feel pleased when seeing others' failures in doing experiments.

related lab exams. The instrument validation process included two stages: In the first stage, I invited three experienced chemistry teachers and four chemistry education researchers to provide suggestions for revising survey items. Following experts' suggestions, I deleted the two items measuring students' emotions in chemistry lab exams (i.e., Q22 and Q23) because some high schools did not have such exams. The item measuring schadenfreude was also deleted due to the concern that students may not express their feelings honestly because schadenfreude has a negative connotation in Chinese. The word "sad" in item 21 was replaced with "disappointed". The words "feel shamed" in item 14 were replaced with "may blame myself" because the latter could represent the meaning of shame and guilty more accurately in the Chinese context. In the second stage, I collected survey data to examine its internal structure. Participants were 939 students from three schools (one middle school and two high schools) located in a northern city in China. The three schools were selected because they represented the student population more broadly: One high school was top-ranked and located in urban areas while the other high school and middle school were located in suburban areas. Among them, 45.3% were female students and 54.7% were male students; 8.4% were ninth grade, 54.8% were tenth grade and 36.7% were 11th-grade. The final administrated survey included 21 Likert-scale items.

Exploratory Factor Analysis

The revised questionnaire, which examined students' achievement emotions in doing chemistry experiments (AEQ-CE), had adequate internal reliability with Cronbach $\alpha = 0.770$. The Kaiser-Meyer-Olkin measure (KMO = 0.875) and Barlett's test of sphericity, $\chi^2 (210) = 7562.650$, $p < 0.001$ indicated that the data were suitable for factor analysis. Before confirmatory factor analysis (CFA), we first conducted exploratory factor analysis (EFA) to examine its psychometric property because these survey items were newly developed. Results from EFA could be used as the evidence of constructing competing models in CFA. Since different types of achievement emotions are assumed to be correlated with each other, I used principal axis factoring with a direct oblimin rotation. Screen plots, eigenvalues (greater than 1) and the pattern matrix suggested that a four-factor solution was tenable, which explained 55.543% of the item variance.

The pattern matrix (see Table 3.6) indicated that item 7 and item 17 showed substantial cross loading while item 11 had weak loading (smaller than 0.3). Other items had strong primary loadings on four factors with the exception of item 3. Therefore, EFA results suggested that item 7, item 11, and item 17 may be excluded from the following analysis. Regarding the four-factor solution, Factor 1 was composed of items describing students' *positive emotions* (i.e., enjoyment, excitement, contentment, and pride) toward chemistry lab classes. Items loading on Factor 2 mainly focused on students' *fear and nervousness*

Table 3.6 The pattern matrix of exploratory factor analysis

Items	F1	F2	F3	F4
1. I look forward to my lab classes.	−.844			
2. I feel excited when I can have chemistry classes in the lab.	−.759			
4. Phenomena observed in chemistry experiments are interesting.	−.825			
5. I feel glad when I can observe obvious phenomena.	−.797			
13. I feel happy when I can work with others to do chemistry experiments.	−.543			
18. I feel satisfied when experiment results are in line with expectations.	−.630			
20. I feel a sense of accomplishment when I can finish the experiment smoothly.	−.646			
3. I feel afraid of taking lab classes because some chemicals are toxic.		−.358		
6. I feel nervous when I do hands-on activities in the lab.		−.637		
7. I feel relaxed when doing chemistry experiments.	−.346	.404		
8. I may feel frightened if some accidents happened in the lab (e.g., breaking the glassware).		−.647		
9. I feel nervous when I am criticized by the chemistry teacher due to improper operations.		−.534		
10. I feel anxious when my experiment progress is slower than others.			.484	
12. I envy that others' experiments have obvious phenomena.			.454	
14. I may blame myself when I hold back the group's progress in doing experiments.			.449	
17. I feel annoyed when I have difficulty analyzing the data.			.384	.329
19. I feel anxious when experimental results are out of expectation.			.706	
21. I may feel disappointed if my experiment did not go well.			.754	
15. I feel angry when the experiment does not go well due to team members' mistakes.				.537
16. I feel bored when waiting for the emergence of phenomena.				.764
11. I feel happy when I can complete experiments quicker than my classmates.				

associated with doing chemistry experiments. The fear may derive from interactions with toxic chemicals or unexpected laboratory accidents while the nervousness may result from teachers' criticism or hands-on activities for operating apparatus. Items loading on Factor 3 described students' *outcome-oriented negative emotions*, which may be represented in different forms including anxiety, envy, shame, and disappointment. The feelings of anxiety and disappointment might result from the risk of failure while envy was related to social comparisons with competitors and shame was related to within-group relationships with cooperators. Items loading on Factor 4 described students' *process-oriented negative emotions* including anger and boredom, which derived from the experimental progress and long wait.

Confirmatory Factor Analysis

Results from EFA provided evidence for constructing multiple-factor models in CFA, which can be used to compare model fitting between different internal structures. Three different models were tested for the AEQ-CE: One-factor model, two-factor model, and four-factor model. The one-factor model included all survey items; items in the two-factor model were grouped into two categories: Positive emotions and negative emotions; and items in the four-factor model were grouped into four categories derived from EFA results: *Positive emotions, fear and nervousness, outcome-oriented negative emotions*, and *process-oriented negative emotions*. Different fit indices including the Standardized Root Mean Square Residual (SRMR), the Root Mean Square Error of Approximation (RMSEA), the Comparative Fit Index (CFI), and Tucker Lewis Index (TLI) were used to assess and compare the fit of the three models. Specifically, SRMR/RMSEA $\leq.06$ and CFI/TLI≥0.90 indicate good model fit (Hu & Bentler, 1999). The fit indices for each model were presented in Table 3.7. CFA revealed that the one-factor model did not fit the data well (SRMR = 0.167; RMSEA = 0.134; CFI = 0.436; TLI = 0.360) while the four-factor model showed considerably better fit (SRMR = 0.057; RMSEA = 0.063; CFI = 0.880; TLI = 0.858). As some items were extremely related and had high modification indices, I revised the four-factor model by adding two covariances (item 1 and item 2, item 18 and item 20). The trimmed four-factor model produced a reasonable fit (SRMR = 0.054; RMSEA = 0.049; CFI = 0.928; TLI = 0.913). All items in the four-factor model had standardized coefficients greater than 0.4 and Cronbach alphas for each factor ranged from 0.661 (i.e., *process-oriented negative emotions*) to 0.879 (i.e., *positive emotions*) (see Table 3.8).

Table 3.7 Fit indices for different models

Model	χ^2	df	TLI	RMSEA	CFI	SRMR
One-Factor	2402.485	135	0.360	0.134	0.436	0.167
Two-Factor	943.874	134	0.770	0.080	0.798	0.085
Four-Factor	609.761	129	0.858	0.063	0.880	0.057
Trimmed Four-Factor	416.482	127	0.913	0.049	0.928	0.054

Table 3.8 Standardized coefficients of four factors

Four Factors	Items	Estimates	S.E.
F1: *Positive Emotions*	Q1	0.757	0.038
(Cronbach α = 0.879)	Q2	0.707	0.038
	Q4	0.814	0.029
	Q5	0.794	0.037
	Q18	0.645	0.049
	Q13	0.579	0.050
	Q20	0.656	0.045
F2: *Fear and Nervousness*	Q3	0.454	0.041
(Cronbach α = 0.718)	Q6	0.611	0.033
	Q8	0.724	0.027
	Q9	0.707	0.029
F3: *Outcome-Oriented Negative Emotions*	Q10	0.643	0.028
(Cronbach α = 0.749)	Q12	0.404	0.034
	Q14	0.536	0.031
	Q19	0.761	0.023
	Q21	0.738	0.023
F4: *Process-Oriented Negative Emotions*	Q15	0.691	0.040
(Cronbach α = 0.661)	Q16	0.715	0.037

Note: all p values < .001.

Inter-correlations between Four Factors

The internal structure of the AEQ-CE was also assessed by examining correlations between four different latent factors (see Table 3.9). Most of the intercorrelations were statistically significant ranging from −0.234 (between *Positive Emotions* and *Fear and Nervousness*) to 0.804 (between *Fear and Nervousness* and *Outcome-Oriented Negative Emotions*). Factors that described students' negative emotions in doing chemistry experiments including Factor 2 (*Fear and Nervousness*), Factor 3 (*Outcome-Oriented Negative Emotions*), and Factor 4 (*Process-Oriented Negative Emotions*) were positively and highly correlated with each other. Correlations between the *positive-emotion* factor and the other three negative-emotion factors were inconsistent: Factor 1 (*Positive Emotions*) was negatively and significantly correlated with Factor 2 (*Fear and Nervousness*) and Factor 4 (*Process-Oriented Negative Emotions*). In contrast, the correlation

Table 3.9 Inter-correlations between factors

Factors	F1	F2	F3	F4
1. Positive Emotions	−	−.234**	−.005	−.374**
2. Fear and Nervousness		−	.804**	.737**
3. Outcome-oriented Negative Emotions			−	.689**
4. Process-oriented Negative Emotions				−

** p < .001.

between Factor 1 (*Positive Emotions*) and Factor 3 (*Outcome-Oriented Negative Emotions*) was very weak and statistically insignificant. These statistical results indicate that correlations between positive emotions and negative emotions may vary depending on sources and types of emotions.

Conclusions and Discussion

Based on the control-value theory, the purpose of the present study was to develop a reliable questionnaire to measure students' achievement emotions about doing chemistry experiments (AEQ-CE) and to examine its internal structure. It adds to the literature by explicating the complex influence of contextual elements in lab environments on students' achievement emotions. The results of this study showed that students' achievement emotions in doing chemistry experiments were diverse and multi-dimensional. EFA revealed a four-dimensional construct: *Positive emotions, fear and nervousness, outcome-oriented negative emotions*, and *process-oriented negative emotions*. CFA supported the four-factor solution and the final instrument had adequate internal reliability with Cronbach α = 0.773. In addition, inter-factor correlations showed that three negative-emotion factors were positively and highly correlated while correlations between the positive-emotion factor and three negative-emotion factors were negative, which aligned well with theoretical assumptions and previous empirical findings (e.g., Artino & Jones, 2012; Pekrun, 2006; Pekrun et al., 2011). All the preceding information provided evidence for the construct validity of the instrument. As existing instruments usually examine students' achievement emotions in regular classroom settings (Pekrun et al., 2005) and only a few studies have examined students' emotional experiences when performing scientific experiments (Itzek-Greulich et al., 2017; Minkley et al., 2017), it remains unknown what and how situational characteristics in the lab setting shape students' achievement emotions. The AEQ-CE instrument can help educational researchers better understand students' achievement emotions across different learning environments. In the following section, I will systematically discuss how four distinct factors relate to contextual factors in lab environments.

Items under the *positive emotion* factor were composed of two types of emotions: Activity-related positive emotions (i.e., enjoyment or happiness derived from conducting chemistry experiments) and outcome-related positive emotions (i.e., satisfaction or pride derived from the successful outcome). It is consistent with the previous finding that students are more interested and motivated to learn science when hands-on experiments are integrated into classroom activities (Haury & Rillero, 1994; Holstermann, Grube, & Bögeholz, 2010; Renner, Abraham, & Birnie, 1985; Thompson & Soyibo, 2002). Therefore, national curriculum standards often recommend that chemistry teachers can adopt hands-on experiments as an effective pedagogical technique for improving student motivation and engagement. Besides its positive influence, the laboratory setting is embedded with unforeseen elements that may contribute to inciting negative emotions.

While positive emotions constituted a single factor, negative emotions were more complicated and differentiated into three distinct factors. The first factor consisted of items describing students' feeling of fear and nervousness when using toxic chemicals and operating laboratory apparatus. These situations are neutral but stressful because students may have to face health or safety concerns and teachers' criticism if unexpected incidents occurred. The feeling of fear, which is related to the appraisal of danger, may lead to actions of escaping the situation (Shuman & Scherer, 2014). Based on the typical sequence of appraisals (Pekrun, 2006), fear and anxiety are produced in the following temporal order: The failure is to be expected and subjectively harmful, but students may be uncertain whether the failure could be avoided, so anxiety is expected to be aroused. The feeling of nervousness or uneasiness is closely associated with anxiety and can be treated as the representation of emotional stress reactions due to the risk of self-attributed failures.

Items under the second factor described students' outcome-related negative emotions including anxiety, envy, disappointment, and shame, which derived from task outcomes and social relationships. Specifically, anxiety and disappointment are assumed to be aroused when negative consequences are expected in an evaluative situation (Zeidner, 1998). Envy usually arises from the context of competition (Brigham, Kelso, Jackson, & Smith, 1997) and shows that students desire the advantage of another (e.g., faster progress in an experiment). Dong and Yu (2007) also listed admiration and envy as commonly experienced emotions among Chinese teenagers though such social emotions were not included in their final survey. Shame is one self-directed or self-conscious emotion and sometimes is intertwined with the feeling of guilt. These two types of emotions are less differentiated in Eastern collectivistic cultures. In the laboratory setting, if one individual impedes the experimental progress during the group collaboration, such an outcome is self-attributed and may influence relationships with others though it may be linked to either controllable or uncontrollable causes (Graham & Taylor, 2014). Negative emotions emerged in this factor have demonstrated that the goal structure of lab learning environments plays an important role in shaping students' achievement emotions. Johnson and Johnson (1975) proposed three types of goal structures based on standards of evaluating achievement: Individualistic structure, competitive structure, and collaborative structure. The lab environments where students perform chemistry experiments in small groups is a hybrid structure including within-group collaboration and between-group competition (Covington, 1992). Different goal structures influence how individuals perceive social relationships with others. Therefore, the complexity of lab environments motivates teachers to develop appropriate standards for evaluating students' lab performance so that the positive effect of doing experiments can be optimized through reaching the balance between collaboration and competition.

Items under the third factor described students' anger and boredom during the process of doing chemistry experiments. The deactivating emotion of boredom is incited because students may feel the activity of long waiting is monotonous and lacks incentive value. Situational antecedents of boredom in the lab setting share similarities with those identified in math classes such as perceived monotony, lack of meaning, opportunity costs, underchallenge, and lack of involvement (Daschmann, Goetz, & Stupnisky, 2011). In contrast, the activating emotion of anger is directed toward others and is assumed to be control-dependent (Pekrun, 2006). In the lab setting, students perceive anger when others are responsible or blamed for undesirable events (e.g., mistaken operations and the failure of experiments) (Kuppens, Van Mechelen, Smits, & De Boeck, 2003). Given the negative influence of boredom and anger on students' motivation, information processing and academic outcomes (Pekrun, Lichtenfeld, Marsh, Murayama, & Goetz, 2017), chemistry teachers may encourage students to reflect on scientific inquiry questions during the waiting period or learn basic emotion regulation strategies including cognitive-approaching coping, behavioral-approach coping, cognitive-avoidance coping, and behavioral-avoidance coping strategies (Nett, Goetz, & Daniels, 2010).

Implications and Limitation

Chemistry lab environments are embedded with rich elements that may incite students' various positive emotions and negative emotions. Though the positive influence of doing chemistry experiments has been systematically discussed in previous research, this study indicates that chemistry education researchers and practitioners should also take students' negative emotions associated with doing chemistry experiments into consideration. For example, chemistry teachers can enhance safety education, model proper ways of using chemical apparatus and provide students with rich opportunities of doing hands-on activities. These pedagogical techniques can help students reduce negative emotions such as fear of using chemical substances and nervousness about doing hands-on activities. What is more, chemistry teachers may set appropriate learning goals and create a friendly atmosphere that strikes a balance between collaboration and competition to foster student learning.

To our knowledge, this is the first pilot study that aims to develop an instrument for measuring students' achievement emotions in doing chemistry experiments. However, several limitations should be considered when interpreting results. First, EFA and CFA results identified a fourth factor (i.e., process-oriented negative emotions) that only included two items, which led to the relatively lower internal reliability of this factor. Future research can use more effective methods to identify other possible negative emotions that occurred in the process of doing experiments (e.g., schadenfreude) or collect data from different populations to reexamine the survey structure. Second, the results of this study showed that the emotion of envy was classified as outcome-oriented negative emotions. In the Chinese context, the valance of such feelings may vary from positive to negative, depending on the degree of envy. On the positive side, it is similar to the feeling of admiration while it is similar to jealousy on the negative side. Future research can explore sources and consequences of the preceding perceptions in more depth. Third, as the instrument was developed for middle school and high school students, it only focuses on the process of doing chemistry experiments and neglects occasions of taking chemistry experiment tests and completing lab reports. As university students may conduct more open-ended and complicated experiments, such environments may require longer waits and thus incite more diverse emotions (i.e., boredom, calmness). More research is needed to develop the instrument targeting university students' achievement emotions in doing chemistry experiments.

Acknowledgements

The author would like to thank Kexin Mao (Tianjin Normal University) who assisted in the data collection.

References

Artino, A. R., & Jones, K. D. (2012). Exploring the complex relations between achievement emotions and self-regulated learning behaviors in online learning. *Internet and Higher Education, 15*(3), 170–175. https://doi.org/10.1016/j.iheduc.2012.01.006

Brigham, N. L., Kelso, K. A., Jackson, M. A., & Smith, R. H. (1997). The roles of invidious comparisons and deservingness in sympathy and pleasure in other's misfortune. *Basic and Applied Social Psychology, 19*, 363–380.

Chi, M. T. (2009). Active-constructive-interactive: A conceptual framework for differentiating learning activities. *Topics in Cognitive Science, 1*(1), 73–105.

Covington, M. (1992). *Making the grade: A self-worth perspective on motivation and school reform.* New York, NY: Cambridge University Press.

Daschmann, E. C., Goetz, T., & Stupnisky, R. H. (2011). Testing the predictors of boredom at school: Development and validation of the precursors to boredom scales. *British Journal of Educational Psychology, 81*, 421–440.

Dong, Y., & Yu, G. (2007). The development and application of Chinese teenagers' achievement emotions. *Acta Psychologica Sinica, 5*, 852–860. (In Chinese).

Graham, S., & Taylor, A. Z. (2014). An attributional approach to emotional life in the classroom. In R. Pekrun, & L. Linnenbrink-Garcia (Eds.). *International handbook of emotions in education* (pp. 106–129). New York, NY; London: Taylor and Francis.

Hareli, S., & Weiner, B. (2002). Social emotions and personality inferences: A scaffold for a new direction in the study of achievement motivation. *Educational Psychologist, 37*(3), 183–193.

Haury, D. L., & Rillero, P. (1994). *Perspectives of hands-on science teaching.* Columbus, OH: ERIC Clearinghouse for Science, Mathematics and Environmental Education.

Hofstein, A., & Lunetta, V. (2003). The laboratory in science education: Foundations for the twenty-first century. *Science Education, 88*, 28–53.

Holstermann, N., Grube, D., & Bögeholz, S. (2010). Hands-on Activities and Their Influence on Students' Interest. *Research in Science Education, 40*(5), 743–757.

Hu, L. T., & Bentler, P. M. (1999). Cutoff criteria for fit indexes in covariance structure analysis: Conventional criteria versus new alternatives. *Structural Equation Modeling, 6*(1), 1–55.

Itzek-Greulich, H., Flunger, B., Vollmer, C., Nagengast, B., Rehm, M., & Trautwein, U. (2017). Effectiveness of lab-work learning environments in and out of school: A cluster randomized study. *Contemporary Educational Psychology*, *48*, 98–115.

Johnson, D. W., & Johnson, R. T. (1975). *Learning together and alone: Cooperation, competition, and Individualization.* Englewood Cliffs, NJ: Prentice-Hall.

Kuppens, P., Van Mechelen, I., Smits, D.J.M., & De Boeck, P. (2003). The appraisal basis of anger: Specificity, necessity, and sufficiency of components. *Emotion, 3*, 254–269.

Mesquita, B., & Karasawa, M. (2002). Different emotional lives. *Cognition & Emotion, 16*(1), 127–141.

Minkley, N., Ringeisen, T., Josek, L. B., & Kärner, T. (2017). Stress and emotions during experiments in biology classes: Does the work setting matter? *Contemporary Educational Psychology, 49*, 238–249.

Nett, U. E., Goetz, T., & Daniels, L. (2010). What to do when feeling bored? Students' strategies for coping with boredom. *Learning and Individual Differences, 20*, 626–638.

Pekrun, R. (2006). The control-value theory of achievement emotions: Assumptions, corollaries, and implications for educational research and practice. *Educational Psychology Review, 18*(4), 315–341.

Pekrun, R., Goetz, T., & Frenzel, A. C. (2005). *Academic Emotions Questionnaire–Mathematics (AEQ-M): User's manual.* Munich, Germany: University of Munich, Department of Psychology.

Pekrun, R., Goetz, T., Frenzel, A. C., Barchfeld, P., & Perry, R. P. (2011). Measuring emotions in students' learning and performance: The Achievement Emotions Questionnaire (AEQ). *Contemporary Educational Psychology, 36*(1), 36–48.

Pekrun, R., Lichtenfeld, S., Marsh, H. W., Murayama, K., & Goetz, T. (2017). Achievement emotions and academic performance: Longitudinal models of reciprocal effects. *Child Development, 88*(5), 1653–1670. https://doi.org/10.1111/cdev.12704

Randler, C., Hummel, E., & Wüst-Ackermann, P. (2013). The influence of perceived disgust on students' motivation and achievement. *International Journal of Science Education, 35*(17), 2839–2856.

Renner, J. W., Abraham, M. R., & Birnie, H. H. (1985). Secondary school students' beliefs about the physics laboratory. *Science Education, 69*, 649–663.

Shuman, V., & Scherer, K. R. (2014). Concepts and structures of emotions. In R. Pekrun & L. Linnenbrink-Garcia (Eds.), *International handbook of emotions in education* (pp. 13–35). New York, NY; London: Taylor and Francis.

Sinatra, G. M., Broughton, S. H., & Lombardi, D. (2014). Emotions in science education. In R. Pekrun & L. Linnenbrink-Garcia (Eds.), *International handbook of emotions in education* (pp. 415–436). New York: Routledge.

Smith, R., Parrott, W., Diener, E., Hoyle, R., & Kim, S. H. (1999). Dispositional envy. *Personality and Social Psychology Bulletin, 25*, 1007–1020.

The Ministry of Education of the People's Republic of China. (2022). *Science Curriculum Standards of compulsory education.* Beijing: Beijing Normal University Press. (In Chinese)

Thompson, J., & Soyibo, K. (2002). Effects of lecture, teacher demonstrations, discussions and practical work on 10th graders' attitudes to chemistry and understanding of electrolysis. *Research in Science & Technological Education, 20*, 25–37.

Zeidner, M. (1998). *Test anxiety. The state of the art.* New York: Plenum.

Part 3

Students' Motivations and Emotions in New Settings

4 How Computer Simulations Influence Students' Motivations and Emotions

Introduction

Computer simulations, defined as computer-generated dynamic models of the real world and processes (Smetana & Bell, 2012), are regarded as one potential and promising approach to transforming teaching and learning in science classrooms. They provide students with new learning opportunities such as interacting with dynamic model systems, visualizing representations of physical phenomena, and receiving animated feedback (Van der Meij & de Jong, 2006). Despite this potential, the effectiveness of using computer simulations is strongly dependent on teachers' pedagogical practices of integration within specific classroom contexts (Hsu & Thomas, 2002; Smetana & Bell, 2012). At present, few studies have explored the feasible pedagogical strategy and corresponding influence of integrating computer simulations into highly structured Chinese classrooms. The present case study aims at filling this research gap by observing how two Chinese teachers implemented computer simulations in their chemistry classrooms and collecting survey and interview data to examine students' affective perceptions in simulation-integrated environments. Such effort can increase the knowledge about expanding the application of computer simulations across classroom contexts.

Pedagogical Strategies and Impacts of Using Computer Simulations in Western Contexts

Computer simulations are now used worldwide in a variety of educational environments such as lecture, laboratory, recitation, homework, and informal settings (Finkelstein, Adams, Keller, Perkins, & Wieman, 2006). Despite its wide application, integrating computer simulations into classroom instruction is a complex undertaking. For example, some teachers are reluctant to make such pedagogical reforms in daily instruction after weighing potential benefits and costs. Classroom teachers, who serve as the main decision maker of the entire process, play critical roles in aligning the use of computer simulations with curricular objectives and student needs in specific classroom contexts

DOI: 10.4324/9781032649665-7

(Hennessy, Deaney, & Ruthven, 2006). Gong, Wei, Bergey, and Shockley (2023) adopted a comparative case study approach to analyze similarities or differences in the two teachers' pedagogical reasoning and decisions of integrating computer simulations in different school contexts. One teacher evaluated affordances and limitations of the computer simulation from a designer's perspective and intended to provide students with careful guidance by constraining system variables while the other teacher reasoned from a learner's perspective and favored more open-ended exploration. Manfra and Hammond (2008) pointed out that teachers' pedagogical aims dominate their pedagogical strategies and perceived values of integrating technology. For example, some science teachers who emphasize content understanding might use computer simulations as a visualization tool to present information. In contrast, other teachers who focus on developing students' scientific practices outlined in the Next Generation Science Standards (NGSS) might use computer simulations as an inquiry tool to perform exploration tasks. How do the two different pedagogical approaches of integrating computer simulations influence students' science learning experiences? To answer this question, I will first review findings of previous research conducted in Western contexts.

Using Computer Simulations as an Inquiry Tool

Computer simulations vary in their degrees of immersion. As some computer simulations can embed science content within highly immersive virtual environments, education researchers recommend using them in a student-centered approach where students can develop scientific inquiry skills and construct conceptual understanding on their own. For example, Ketelhut (2007) and her colleagues implemented an interactive simulation – *River City* – for engaging middle school students in collaborative scientific inquiry and developing 21st-century skills over three weeks. Students visited virtual environments six times to familiarize themselves with the interface, complete mini-tasks, and test hypotheses. The results showed that the problem-solving process of engaging in virtual experimentation increased academic self-efficacy of students in the experiment group (Ketelhut & Nelson, 2010). This conclusion is further supported by qualitative interview data: Students who found the regular science class boring or had low feelings of self-efficacy were reported to persistently figure out the presented problem in virtual environments and enjoy the science class more (Clarke & Dede, 2005). Accordingly, computer simulations are suggested to effectively engage middle school students in learning science and "act as a catalyst for change in students' self-efficacy" (Ketelhut, 2007, p. 99). Meluso, Zheng, Spires, and Lester (2012) corroborated the above argument with the evidence that fifth-grade students' science self-efficacy and science content learning significantly increased after interfacing with a simulation microworld called *Crystal Island* across a series of four days. In their study, students completed an online tutorial to get familiar with the controls and character movements on the first day, then interacted with the computer simulation either on a single-player condition or a collaborative playing condition for 40–50 minutes on the following three days. Both studies indicate that computer simulations can potentially increase students' self-efficacy, positive emotions (e.g., enjoyment), and engagement in learning science.

Despite these benefits, it is difficult to generalize the student-centered pedagogical approach to specific science classrooms where class time and technology equipment are limited. To solve this problem, other researchers have explored the possibility and influence of demonstrating computer simulations in a teacher-directed approach. It should be noted that such a pedagogical approach does not necessarily prohibit inquiry opportunities. Instead, computer simulations can be useful tools for interactive lecture demonstrations, which support whole-class inquiry practices (McKagan et al., 2008). For example, Rutten, van der Veen, and van Joolingen (2015) systematically described how one Dutch secondary school physics teacher integrated computer simulations to support the "predict-observe-explain" cycle (Hennessy et al., 2007), which is one important principle in the majority of inquiry approaches (Bell, Urhahne, Schanze, & Ploetzner, 2010). Specifically, students first predicted how scientific phenomena would develop, then observed and described the actual phenomena, and finally explained why phenomena developed in certain ways. The results showed that using computer simulations as an interactive demonstration tool enhanced students' attention focus, enjoyment, interest, and science knowledge. However, researchers also reported that students in large classes were less convinced that teaching with computer simulations contributed to their motivation. In other words, the large class size might counteract the positive effect of computer simulations on students' affective perceptions. Such an issue provides the rationale of conducting the current study: If computer simulations are integrated into large Chinese high school classrooms, what is the influence on students' affective perceptions?

Using Computer Simulations as a Visualization Tool

Using computer simulations as an inquiry tool requires science teachers to reform their regular teaching strategies. Due to various contextual challenges and pedagogical aims, many science teachers are reluctant to adopt the inquiry-based approach and prefer to integrate computer simulations as a visualization tool. In this situation, students are asked to observe the demonstration and provide or listen to explanations for scientific phenomena. There are two different ways of using computer simulations as a visualization tool. Some science teachers use computer simulations as an alternative to traditional textbook-based instruction. For example, Kiboss, Ndirangu, and Wekesa (2004) replaced textbooks with computer-mediated simulations in the secondary biology course over a three-week period. During these lessons, students were presented with animated color graphics and short notes with factual information on cell division. The results showed the experiment group reported significantly higher gains in positive perceptions of classroom environments and feelings toward the biology course than the control group. Such positive effects were attributed to students' active interactions with the simulation, which simplified "mystic" concepts in science discourse. Other teachers use computer simulations to supplement the traditional classroom instruction. For example, Jimoyiannis and Komis (2001) examined the effects of one computer simulation – *Interactive Physics* – on high school students' understanding of basic kinematical concepts concerning simple motions through the Earth's gravitational field. In the computer lab, the physics teacher used the computer simulation to display simple kinematical phenomena and analyze the free fall laws. The results showed that students in the experiment group exhibited significant improvement in achievement for the tasks concerning the concept of acceleration. Similarly, Stern, Barnea, and Shauli (2008) investigated how a dynamic software simulation – *A Journey to the World of Particles* – influenced middle school students' conceptual understanding of the kinetic molecular theory. The experiment group, who observed the consequences (e.g., traces of an individual particle) of modifying parameters (e.g., temperature and pressure), scored significantly higher than students in the control group. However, the average performance of both groups was low and long-term learning differences were negligible.

In summary, most existing studies that examined effects of using computer simulations as a visualization tool have focused on students' content understanding while ignoring their affective perceptions. It is unknown whether such a pedagogical approach is also productive for improving students' affective perceptions (e.g., self-efficacy, positive emotions, and engagement in learning science), especially in large Chinese classrooms. By answering this question, this study can provide empirical evidence for informing the decision making of integrating computer simulations within similar classroom contexts. As mentioned earlier, any discussion about the positive effects of using computer simulations should be accompanied by a discussion about teachers' pedagogical strategies of integration within specific classroom contexts. Before exploring whether such positive effects are still applicable in Chinese large chemistry classrooms, I will first describe how Chinese teachers implemented computer simulations as a visualization tool.

Using Computer Simulations as a Visualization Tool in Eastern Contexts

Many studies that examine learning effects of computer simulations are conducted in experimental conditions and disregard the ecological validity of real classroom environments. The current study aims to fill the research gap by avoiding interfering with teachers' practices of using computer simulations. In this study, I introduced one specific computer simulation called "Reaction & Rate" (Figure 4.1) from the Physics Education Technology (PhET) project, which covers content knowledge about the reaction rate, related influencing factors, and chemical equilibrium (https://phet.colorado.edu/en/simulation/reactions-and-rates). It should be noted that PhET simulations are different from game-based simulations in the interface design. The former allows students to adjust variables and observe dynamic animations while in the latter students can manipulate the character to navigate within the 3D immersive virtual environments.

Graphic Section

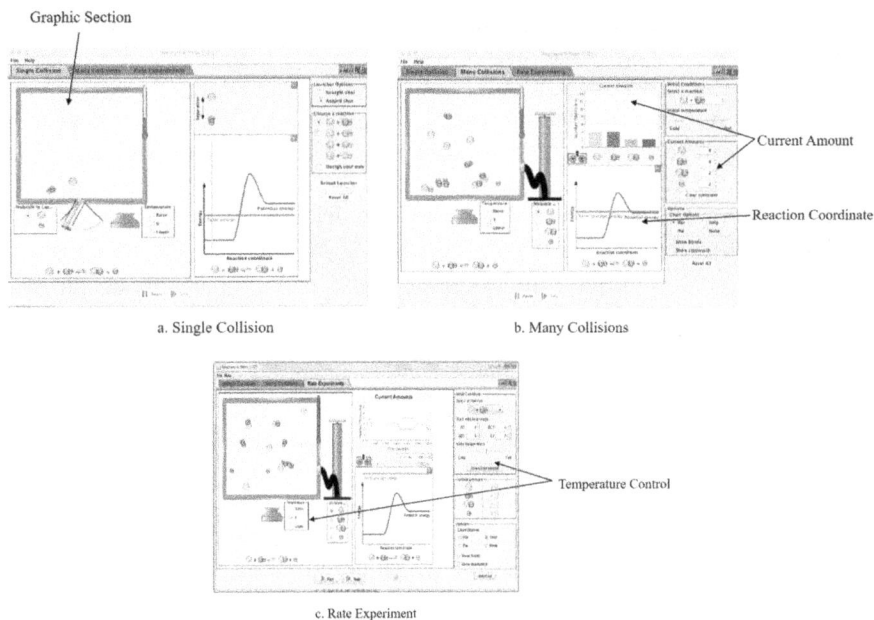

a. Single Collision   b. Many Collisions

Current Amount

Reaction Coordinate

Temperature Control

c. Rate Experiment

Figure 4.1 The screenshots of three panels in the PhET simulation developed by University of Colorado Boulder, licensed under CC-BY-4.0.

Before the classroom implementation, the two Chinese teachers first explored the PhET simulation and evaluated how different simulation features functioned to represent curriculum contents. Taking the richness of simulation information and the realities of classroom contexts (e.g., large class size, heavy curriculum task, limited class time, and technology equipment) into consideration, they proposed that the feasible and effective strategy of integrating computer simulations was to use them as a visualization tool to review related content knowledge in one class period. Such instruction decision is consistent with the Western phenomenon that science teachers are more likely to initiatively integrate computer simulations as a visualization tool. Specifically, the teachers adjusted the system variables (e.g., temperature, concentration) in the computer simulation while students observed and connected animated phenomena with curriculum knowledge. It should be noted that the computer simulation was not integrated to replace traditional experiments that demonstrated chemical phenomena at the macroscopic level (e.g., how water bath heating influences the decomposition rate of hydrogen peroxide). Instead, they were used to supplement the traditional chemistry instruction and elaborated the molecular movement at the microscopic level. The process of using the simulation in the two chemistry classes was videotaped.

Detailed descriptions of what happened in these Chinese chemistry classrooms can help researchers and practitioners comprehensively understand teachers' pedagogical strategies of integrating the computer simulation, and how teachers interacted with students in real classroom contexts. Since the two teachers cooperated to make instructional decisions of integrating the computer simulation, their pedagogical strategies were similar. In this section, I transcribed one of the videotapes as an example and narrated the classroom story of how the chemistry teacher implemented the computer simulation to review related content knowledge. During the simulation-integrated chemistry class, the conversation between the chemistry teacher and students occurred in the format of choral responding. All dialogues were translated from Chinese.

At the beginning of the simulation-integrated class, the chemistry teacher first reviewed the topic (i.e., the reaction rate and influencing factors) by asking a series of questions. The teacher first asked, "What physical variables can influence the reaction rate?" Students responded with different answers such as surface area, temperature, pressure, concentration, and catalyst. Then the teacher connected curriculum contents with industrial examples and asked, "What can be done to increase the surface area of ore in industrial production?" Students provided different methods such as crush, dissolution, and stir. Subsequently, the teacher started to help students review how other physical variables influenced the reaction rate, which could be demonstrated with the computer simulation.

Teacher: How do temperature, pressure, and concentration influence the reaction rate?

Students: (Some students) The pressure influences the distance between molecules or atoms. (Other students) The concentration influences the number of molecules per unit volume.

At this moment, the teacher displayed the interface of the computer simulation on the projector. Students broke into cheers and said "Wow". The teacher then asked some guiding questions and explained simulation features before the demonstration, which are recommended practices of using computer simulations (Hsu & Thomas, 2002).

Teacher: The occurrence of chemical reactions requires molecules to collide [with each other]. How do different factors influence the reaction rate through microscopic collisions? How does the temperature influence the reaction rate?

(Pointing to the chemical reaction at the bottom of the graphic section) Here is the example: If A reacts with BC, [which] consists of B element and C element, and produces AB and C. [This is] one simple replacement reaction. How does temperature influence the reaction rate? Let's add 30A and 30BC.

(Pointing to the graphic section) Here we can see the container is fixed. We can change the number of A and BC to change the concentration, right?

(Pointing to the slider) We can find the "initial temperature" [here]. We can adjust to show whether the temperature is high or low.

(Pointing to the graphic section again) We can also increase or decrease the temperature at the bottom of the container during the reaction process.

(Pointing to the right side of the interface) We can control the concentration here. The initial temperature is here. Let's observe how molecules collide to react.

When the teacher clicked the "start" button, students started to observe molecule movements in the container and wondered with the sound of "wow" again. Simultaneously, the teacher elaborated on how students could connect between different sections of the computer simulation to generate a relationship between temperature and reaction rate.

Teacher: The yellow one is A. The purple and gray one is BC. If we want to see clearer whether new substances are produced, (Pointing to the diagram section) we can observe the [current amounts] coordinate, which shows current molecular amounts of A and BC. We can see [the numbers of] products are increasing. So how does temperature influence [the reaction rate]? Now let's increase the temperature and observe how it influences the reaction rate. Is the reaction rate increased?

Student: Yes.

As the temperature increased, container molecules gradually moved very fast and students started to laugh. One student said to her neighbor, "These balls are flying." Then the teacher guided students to connect the observed phenomenon with the "reaction coordinate".

Teacher: Have you noticed any changes in the "reaction coordinate" when I increase the temperature?

Students: The total energy increases.

Teacher: So how does temperature influence the reaction rate?

Students responded to the question with various answers. The teacher summarized students' ideas,

If the temperature increases, molecules move faster and this increases the collision frequency. We can see the energy line also increases. It shows the [total] reactant energy of A and BC. What is the result? More general molecules become activated molecules. Then [higher temperature] increases the percentage of activated molecules and the final reaction rate.

After clarifying the microscopic mechanism of how temperature influenced the reaction rate, the chemistry teacher moved to the third factor – concentration – and started to demonstrate the second experiment. In this process, she followed students' strategy of increasing the concentration and asked students to observe and compare the reaction rate. In addition, she connected different factors and explained that the fourth factor – pressure – influenced the reaction rate by changing the concentration.

Teacher: Let's observe the concentration. Now we have 30A and 30BC. The container volume is fixed. How can we change the concentration?
Students: (One student) Press the container. (Other students) Add more A and BC.
Teacher: How many do we add?
Students: 100A.
Teacher: (Following the suggestion) OK. Let's increase the concentration of A, and add 100A.

As the container was full of molecules, students were excited, with loud acclamation of "yay". Simultaneously, the teacher guided students to generate a relationship between concentration and reaction rate. Based on students' explanations, she provided further information and used the energy line in the "reaction coordinate" to correct students' misunderstandings.

Teacher: There are so many A. They cannot avoid colliding with BC. If BC moves around, they meet A. Then it is quite possible to react. So how does the concentration influence the rate?
Students: (Some students) Change the number of molecules in per unit volume. (Other students) Change the number of activated molecules.
Teacher: (Pointed to the reaction coordinate) Are there any changes in the energy diagram?
Students: No.
Teacher: We did not change the number of activated molecules. Only increased the number of molecules.

After demonstrating two experiments using the computer simulation, the teacher asked how the last factor – the catalyst – influenced the reaction rate. Students responded with the answer, "Change the activation energy". One student murmured, "Let me try it". The teacher commented, "The computer simulation does not include the function of adding catalyst. It should be improved in the future".

Research Question

How does one PhET simulation influence students' affective perceptions (i.e., chemistry self-efficacy, achievement emotions, and classroom engagement) when it is integrated as a visualization tool for reviewing content knowledge in a traditional Chinese classroom setting?

Methodology

Procedure and Data Collection

The study was conducted in an exemplary high school located in the capital city of a northern province in China. According to student performance in college entrance examinations, this two-campus high school ranked fourth in the located city. Participants were 103 16 or 17-year-old 11th-grade students (45 female and 58 male) from two chemistry classes in the same high school. They were from families with middle socioeconomic status. As the two classes were equivalent in average performance, the classroom effect was ignored and participants were treated as a single sample.

Measures

Students first responded to a questionnaire that consisted of three measures (i.e., chemistry self-efficacy, achievement emotions, and classroom engagement) with respect to regular chemistry classes. All responses were indicated using a 5-point Likert scale, anchored at 1 (strongly disagree) and 5 (strongly agree). The original Chinese language version of Achievement Emotion Questionnaire – Mathematics (AEQ-M; Pekrun, Goetz, & Frenzel, 2005) measures students' emotions during three occasions: Attending classes, doing homework, and taking exams. In this study, I focused on the situation of attending chemistry classes and used the subset of the questionnaire (see Appendix in Chapter 2). The validation has been conducted for the German and Chinese language comparison. Survey items that measured chemistry self-efficacy beliefs and classroom engagement were translated from English to Chinese by the author, and then back-translated by one chemistry teacher, who verified the match with the original version. After the classroom implementation of the PhET simulation, students filled out the post-survey, which included the same items as the pre-survey, to report their perceptions in the simulation-integrated session. Almost all students from the two classrooms filled out pre- (N = 103) and post-surveys (N = 101). The number of items and Cronbach's alphas of pre- and post-surveys for three variables are listed in Table 4.1.

In addition, the two teachers and nine of the students volunteered to participate in semi-structured interviews, which asked questions about their perceptions of using PhET simulations in chemistry classes (Appendix 1). All interviews were audiotaped: The two teachers were interviewed together while the students were interviewed individually in a private office. Sixty-one students (nearly half of the sample), including six of the students who were interviewed, also volunteered to respond to eight open-response questions, which asked about their learning experiences in simulation-integrated classroom environments (Appendix 2). Two example interview and open-response questions were: "How do you feel about simulation-based chemistry class? What is the influence of using the computer simulation on your perceived confidence in learning chemistry?"

Table 4.1 Measures, psychometrics, and example items

Measures and Sources	N	Example Items	Cronbach's alphas	
			Pre-	Post-
Chemistry self-efficacy (Dalgety, Coll, & Jones, 2003)	16	I can convert the data obtained in a chemistry experiment into a result.	.942	.965
Achievement Emotions (Pekrun, Goetz, & Frenzel, 2005)	14	I look forward to my chemistry classes.	.794	.794
Classroom engagement (Skinner, Furrer, Marchand, & Kindermann, 2008)	10	When I'm in chemistry class, I think about other things [reverse coded].	.896	.925

Analytic Plan

Since the study was exploratory in nature and examined students' various affective perceptions in real classroom contexts, I adopted a case study approach to answer the research question. First, such an approach can be of value where the research aims to investigate a complex phenomenon embedded in the real world, and where the scope is difficult to define and can only be understood within context (Yin, 2003). Second, the case study is suitable to uncover interactions of inseparable variables that are elements of the phenomena being studied (Yin, 2003). The quantitative and qualitative data were collected from four sources: 5-point Likert scale surveys (see Appendix in Chapter 2), classroom observations, open responses (Appendix 1), and teacher/student interviews (Appendix 2). The multiplicity of data sources revealed the complex phenomenon from different perspectives and strengthened the convincement of results.

As students may describe various feelings in interviews and open responses, I followed the most commonly used strategy and categorized achievement emotions scale items into six positive emotion items (α = .843) and eight negative emotion items (α = .736). Then I conducted independent sample t-tests to compare means of the four affective variables in the pre- and post-surveys because student identification information was not available for dependent t tests. Regarding qualitative data, I identified units from interview quotes describing how and why computer simulations influenced four affective perceptions (open coding), and finally grouped these related units under different categories (axial coding) (Corbin & Strauss, 1990). My interpretations of the impact of the computer simulation are presented next.

Results and Discussion

The purpose of the present study was to examine Chinese students' affective perceptions when a PhET simulation was integrated as a visualization tool to review content knowledge. Such effort increases the knowledge about how to

Table 4.2 Four variables in lecture-based and simulation-integrated instruction

Variables	Pre-survey (N = 103)		Post-survey (N = 101)		Independent t tests (df = 202)	
	M	SD	M	SD	t	p
Chemistry Self-Efficacy	3.32	.65	3.56	.77	−2.38	.018
Positive Emotions	3.56	.59	3.81	.63	−2.94	.004
Negative Emotions	2.39	.53	2.35	.61	.445	.657
Classroom Engagement	3.67	.68	3.76	.74	−.839	.403

improve students' science learning experiences in Chinese classroom contexts through integrating computer simulations. In this study, the quantitative survey data outlined the overall impact of the computer simulation on students' four affective variables (i.e., chemistry self-efficacy beliefs, positive emotions, negative emotions, and classroom engagement) (see Table 4.2) and qualitative data provided more details about individuals' diverse perspectives behind the scene. Interview quotes that are related to four affective variables are listed in the third column of Table 4.3 with different numbers (e.g., S1, S2... S14) and other quotes are embedded in paragraphs with brackets. In the following sections, I will discuss the effects of the computer simulation along each affective variable.

Chemistry Self-Efficacy Beliefs

The first conclusion is about how the computer simulation influenced students' chemistry self-efficacy beliefs. The independent samples t tests showed that the PhET simulation significantly and positively increased students' chemistry self-efficacy beliefs (t [202] = −2.38, p = .018) (see Table 4.2). Based on Cohen's (1988) rules of thumb, the effect size for this analysis (d = .34) was small to medium. Such significant results reflected the dynamic and malleable property of self-efficacy beliefs, which can be experimentally augmented in a short period of time upon receiving contextual information (Bong & Skaalvik, 2003). Among 61 open responses to Q5 in Appendix 2, 23 students (37.7%) reported a higher level of confidence in learning chemistry (e.g., S1, S2, and S3 in Table 4.3) while 29 students (47.5%) reported a lack of a big influence (e.g., S4, S5, and S6 in Table 4.3). In other words, the survey data indicated an overall positive effect while qualitative data implied that the change in chemistry self-efficacy beliefs was perceived differently based on students' interpretations of the results of simulation-integrated classroom activities.

Why did some students perceive higher chemistry self-efficacy beliefs in simulation-integrated chemistry classes? When answering this question, 15 students ascribed to the power of technology in representing microscopic scientific

Table 4.3 The summary of qualitative data

Variables	Open Responses (N = 61)	Student Quotes
Chemistry Self-Efficacy	"Increased" (N = 23)	S1: "It makes abstract concepts more vivid and easy to understand and grasp. This is the first time I have the idea that chemistry is so easy".
		S2: "It helps. The concrete format conforms to my way of thinking and promotes understanding".
		S3: "It helps to some extent. [I am] clearer about various reaction processes. These concepts are no longer vague. But it does not help much with exams. [It] might be useful when learning new content".
	"No [big] influence" (N = 29)	S4: "No influence at this moment. It does not help much with my knowledge because I have learned them. But it helps deepen my understanding of knowledge".
		S5: "No big influence. I think basic knowledge and test scores influence confidence. Such software only makes classes interesting but does not benefit much for preparing exams".
		S6: "It influences my learning ability rather than confidence".
		S7: "It can help [me] learn more intuitively. [It has] no relations with my confidence. The confidence is an inborn feeling".
Positive Emotions	"Excited" (N = 37)	S11: "It makes us enjoy [chemistry] classes more and facilitates the discussion".
	"Happy" (N = 27)	S12: "My feelings change from nervousness to excitement. It is a challenge for me to take
	"Curious" (N = 25)	chemistry classes, which brings pressure and tension. The computer simulation is
	"Enjoyable" (N = 19)	more like one type of game, which can help relieve my tension".
Negative Emotions	"Less boring" (N =1)	S13: "I am very anxious. It is a waste of time taking this class. I would rather do more
	"Less drowsy" (N =1)	exercises".

Classroom Engagement	"More engaged" (N = 42)	S14: "I can engage more. Very effective with teacher's explanations".
	"Complex" (N = 6)	S15: "The computer simulation helps me concentrate for a while. But the concentration might come from the novelty [of technology], which might distract our attention to the content knowledge itself".
		S16: "The curiosity helps me engage more but sometimes I only observe the screen and neglect the teaching content".
		S17: "I am superficially engaged and watch it for entertainment. I have not learned anything new".
	"Less engaged" (N = 2)	S18: "Partly, it distracts me. Perhaps we cannot devote ourselves to what we are doing when we are too excited".
		S19: "I cannot engage because the classroom order decreases, and I cannot hear what the teacher says".

processes in a vivid manner, which reduced the difficulty of understanding the same content (e.g., S1 and S2 in Table 4.3). Such an explanation is consistent with Bandura's (1997) theoretical assumption that students' self-efficacy beliefs are reciprocally and recursively related to cognition. The progress in cognitive understanding prompts students to perceive higher levels of self-efficacy beliefs. Even though students did not have computers at their disposal, the whole-class demonstration facilitated them to effectively connect between visualized animations and content knowledge under teacher guidance. These colored animations, which provided additional situational resources through visualizing unobservable microscopic phenomena (e.g., S8), reduced students' cognitive loads and allowed them to develop an intuitive understanding of how scientific processes operated and attain a sense of familiarity (Laurillard, 1992). The unique affordances of the computer simulation effectively complemented traditional lectures and potentially benefited students who were disadvantaged in classroom environments with pure linguistic descriptions (e.g., S9 and S10). Interview quotes from three students (S8, S9, and S10) are listed below:

> S8: "It gives life to molecules, I feel more relaxed and no longer have the feeling of seeing the world in the smoke and mirrors".
>
> S9: "I can intuitively observe the reaction process. In the old lecture-based class, I can only figure it out on my own. Sometimes it is difficult to imagine how the [molecular] movement becomes faster or slower".
>
> S10: "Computer simulations can help those who want to learn and spend effort but cannot learn. If the difficulty is decreased, I can understand the same knowledge more easily. For example, I can understand 30% [of content] in old classes but now [I can] understand 50% and get more information".

Besides cognition processes, interview data indicated that the sense of self-efficacy beliefs was associated with students' goal orientations unexpectedly. Individuals hold either a learning goal orientation or a performance goal toward academic tasks (Dweck, 1986). A learning goal orientation describes a desire to master materials and enhance competence or knowledge. In contrast, a performance goal orientation reflects a desire to maximize favorable evaluations of competence (Wigfield & Cambria, 2010). Previous research has shown that learning goal orientation was positively related to self-efficacy, whereas performance goal orientation was negatively or not related to self-efficacy on an academic task (Bell & Kozlowski, 2002; Phillips & Gully, 1997).

The difference in goal orientations might provide explanations for students' different opinions about the effect on chemistry self-efficacy beliefs. In this study, the computer simulation was integrated to review previously learned knowledge. Students with a learning goal orientation (e.g., S1 and S2) valued the computer simulation's function of deepening conceptual understanding and maintaining long-term memory. As one student said,

The computer simulation can help our memory last longer. Teachers' words are easily forgotten. But after using the computer simulation, maybe I cannot recall what the teacher said after two weeks, but the image lingers in my mind. It reminds me of what happened at that time.

One chemistry teacher also supported the computer simulation's role of promoting long-term conceptual understanding because students could apply similar mechanistic knowledge in different contexts:

I think the good thing is that abstract things are more intuitive. The software can represent microscopic formats. Though students might collect less information at the moment, they can extend [in the future]. Chemistry is from the microscopic to the macroscopic level, then from the macroscopic to the microscopic level. If [students are] clear about microscopic things, then they can easily understand the macroscopic level. If they thoroughly understand this microscopic thing, then the other microscopic thing is easy. For example, we can say that the concentration increases collisions. We can say temperature also increases collisions. If I teach [how] the consternation [influences collisions] clearly, then [how] the temperature [influences collisions] is easy to explain. I think this is the advantage. This is like giving [students] one example, they draw other inferences.

In contrast, students with a performance goal orientation (e.g., S4 and S5) disvalued the computer simulation's benefits of promoting deeper understandings of scientific mechanisms or processes because they cared more about performing well in standardized tests, which included questions about a series of factual information. The failure of acquiring new knowledge information might interfere with students' perception of the change in chemistry self-efficacy beliefs (e.g., S3, S4, and S5). Even though the computer simulation had deepened students' understanding of mechanisms between influencing factors and reaction rate, whether it is counted as worthwhile knowledge is a question. In other words, students' deeper understanding might not be transformed into high test scores in school standardized assessments. This argument is supported by the evidence that the performance of the experiment group who used computer simulations was lower than the control group one year after the instruction (Stern, Barnea, & Shauli, 2008). Especially in Chinese highly structured classroom environments with the pressure of college entrance examination, student academic performance is the most important factor for principals, teachers, parents, and students to make academic decisions. Such value norms prompt students to treat the attainment of chemistry knowledge and science performance as the main source of self-efficacy beliefs.

In addition, some students' conclusions about the lack of change might result from their personal philosophy about the malleability or stability of chemistry self-efficacy beliefs. When responding to Q5 in Appendix 2, two students argued that the computer simulation contributed to improving perceived abilities rather than the perceived confidence in learning chemistry. They argued that the confidence was inborn and more stable than the ability (e.g., S6 and S7). Therefore, chemistry self-efficacy beliefs should be developed and shaped over longer periods. As one student commented in the interview, "It is too early to conclude" because the computer simulation in this study was integrated for only one class period. However, students' viewpoints are quite different from researchers' argument that the ability "may be changeable, but only after a long period of time" (Gist & Mitchell, 1992).

Positive Emotions

The second conclusion is about how the computer simulation influenced students' perceptions of positive emotions. Pre- and post-survey data indicated that the computer simulation significantly and positively increased students' positive emotions (t [202] = -2.04, $p = .004$) (see Table 4.2), which is consistent with the results of previous research (Kiboss, Ndirangu, & Wekesa, 2004). Based on Cohen's (1988) rules of thumb, the effect size for this analysis ($d = .41$) was medium. Such a positive effect was further supported by 61 open responses to Q1, Q2, and Q4 in Appendix 2. When using various terms to summarize their general feelings about simulation-integrated chemistry classes, 37 students described their emotions with the term "excited", 27 students with "happy", 25 students with "curious", and 19 students with "enjoyable".

How did external classroom environments shape students' perceptions of positive emotions? Pekrun, Goetz, Titz, and Perry (2002) noted that positive emotions are developed based on two different types of cognition: Thoughts about the learning materials and appraisals of mastery and success. Specifically in simulation-integrated chemistry classrooms, the novelty of learning materials (i.e., computer simulation), defined as the perceived newness of an innovation, might have stimulated individuals' positive affective reactions because the integration of such innovative technology tool represented a significant improvement over its existing predecessor (Wells, Campbell, Valacich, & Featherman, 2010). Such an argument is supported by qualitative data. Among 61 open responses, 27 students attributed their positive emotions (e.g., curiosity, excitement) to experiencing interactive technology in a chemistry class for the first time. They compared with previous learning experiences and suggested that the integration of computer simulation transformed the traditional lecture-based instruction (e.g., "using computers is better than using blackboards") and made chemistry class more enjoyable (e.g., S11 and S12 in Table 4.3). Students' emotional states of excitement were also reflected in their sounds of "wow" and "yay" when the chemistry teacher operated simulation variables. As one teacher said in the interview, "Students see new things. They are excited when new animations are presented. Such excitement can infect classmates around and increase the [classroom] discussion or extend [students'] imaginations". Someone might be concerned about the sustainability of positive emotions. In other words, curiosity and excitement from computer simulations may be temporary and may disappear if they were integrated in the long term. The qualitative data indicated the answer might be "no". As one student commented, "If the teacher uses this computer simulation again, she will use other functions. They are different from [the function or content knowledge] this time. If the teacher uses a different computer simulation next time, these features are also different".

Regarding the second factor, students' sense of being able to master the material is another main source of students' positive emotions (Pekrun et al., 2002). For example, one student commented, "I am happy when I understand (the content)". Students' perceptions of successfully understanding content knowledge, which is indicated by increased chemistry self-efficacy beliefs, create conditions for perceiving positive emotions. In Chinese highly structured classroom environments, students' positive emotions from understanding the knowledge might be more valuable than those from the simulation itself. Therefore, the advantage of improving students' positive emotions would still exist as long as computer simulations could supplement traditional lecture-based chemistry instruction and promote conceptual understanding. Just as one teacher commended in the interview:

The most impressive part of using the simulation is (to show) increasing the concentration of A can increase the reaction rate. [Usually] I orally describe that the number of molecule A in per unit volume increases so the reaction rate increases naturally. But how does this process happen? Actually, I do not point it out, or it is difficult to describe clearly with oral language. If using the animation, students can see B is surrounded by A. It is difficult to avoid the collision and reaction. As the [frequency of] collision increases, the percentage of reacting also increases".

The simulation's affordances of visualizing microscopic processes to promote intuitive understanding are particularly evident in chemistry where students often need to construct different mental models for explaining and understanding scientific mechanisms. In many cases, students are uncertain or confused about whether their imagined assumptions are correct or not. The diverse representations in the computer simulation can enrich students' experiences of learning scientific concepts. As one student noted,

I was confused when I learned the content. I carefully read [textbook content] and thought about what was correct. Then I followed [my way of] thinking and memorized it at that time. Today I observe the simulation and realize that it should be like this. It breaks my old thinking.

These visualized dynamic animations, which can be used to evaluate the validity of mental models, might contribute to bridging the cognitive gap between texts in lectures and images in minds. As indicated in the teacher interview:

> "It has the function of guidance. Some chemistry knowledge like the inner crystal structure is so difficult for students who have limited [abilities of] spatial imagination or [students who] have never been exposed to such things. If they could not imagine [the situation], then much work could not continue. ...In some cases, two people are talking and describing [the same thing]. For instance, the model should be in the format of A. But the imagination is B. A and B have something in common. Our verbal description may sound the same, or cannot show the difference. In fact, they are different".

Negative Emotions

The third conclusion is about how the computer simulation influenced negative emotions. Despite the fact that the current study extends prior literature by taking students' negative emotions into consideration, the mean difference in negative emotions in pre-and post- survey was minimal (t [202] = .445, p = .657) (see Table 4.2). The effect size for this analysis (d = .07) was negligible. In other words, the integration of computer simulations had a negligible effect on students' negative emotions. Open response data also showed that students were more likely to use positive terms (e.g., interested, excited, happy) rather than negative ones (e.g., angry, anxious) to describe their feelings. Only four students mentioned negative emotions at all. Among them, three students said that computer simulations "relieve the tension" (e.g., S12) and made chemistry classes "less boring" and "less drowsy" while one student said that he was anxious because "it is a waste of time taking this class. I would rather do more exercises" (e.g., S13). Such phenomenon is consistent with the previous finding that students reported positive emotions more often than negative emotions concerning situations of attending classes (Pekrun et al., 2002). One possible explanation is that negative emotions such as anxiety and anger are usually outcome directed while this study focused more on students' emotional experiences related to classroom activities. For example, the feeling of anxiety might result from the worry about the failure of improving science performance. In addition, students may be discouraged or unaccustomed to expressing negative emotions in Chinese classroom climates. Gong and Bergey (2020) have systematically described how Eastern cultural norms shape students' emotional expressions.

Classroom Engagement

The fourth conclusion is about how the computer simulation influenced student engagement in chemistry classrooms. The pre- and post-survey data indicated the computer simulation did not significantly increase the level of engagement (t [202] $= -.839$, $p = .403$) (see Table 4.2). The effect size for this analysis ($d = .11$) was small. Among 61 open responses to Question 6 in Appendix 2, 42 students reported that they were more engaged in simulation-integrated chemistry classes (e.g., S14 in Table 4.3). Six students held more complex opinions (e.g., S15, S16, and S17 in Table 4.3). On the one hand, the computer simulation attracted them to observe the dynamic movement of animated balls on the screen. On the other hand, the sole concentration on the graphic section might lead to the neglect of teacher guidance and explanations, which might distract students from connecting the computer simulation with the content knowledge. Two students thought that the computer simulation distracted their attention due to the over-excitement toward the computer simulation and the decreased classroom order (e.g., S18 and S19 in Table 4.3).

The inconsistency between the survey data and qualitative data might result from the multidimensional nature of classroom engagement (Fredricks, Blumenfeld, & Paris, 2004), which is reflected in survey items and interview quotes. The survey items evaluated students' engagement based on cognitive activities such as the process of using the computer simulation to achieve conceptual understandings. For example, two example survey items were "When I'm in chemistry class, I listen very carefully" and "When I'm in chemistry class, my mind wanders". In contrast, qualitative data provided engagement information based on external classroom behaviors. For example, students reported that they were attracted by the new learning material specifically the graphic section of the simulation interface, which presented molecular collisions in a dynamic approach. However, students explained that the full attention to the molecular movements resulted from their curiosity in using the computer simulation for the first time. If computer simulations were integrated over the long term, the novelty effect might wear off and students might be more cognitively engaged due to the awareness of being responsible for their own learning. As one student commented, "The class time is short. These things are secondary. [I will] listen to teachers first. If teachers emphasize such issues, the problem [of distraction] can be avoided". This commentary also highlights the importance of teacher guidance when using computer simulations, especially in large classes, which is consistent with previous findings (Smetana & Bell, 2012). Such guidance might include giving hints about where to observe and asking guiding questions about how different sections are related to each other. The teacher's guidance might help monitor students to transform behavioral engagement into cognitive engagement that connects the observed dynamic animations with related content knowledge.

Results showed that the relationships among affective perceptions might be more complicated in simulation-integrated classroom environments. Even though there were no classroom effects for pre-survey, preliminary multivariate analyses of variance (MANOVAs) of post-survey revealed that there were significant differences between the two classrooms with respect to the four affective variables, $F (4, 96) = 4.269$, $p = .003 < .05$; Wilk's $\Lambda = .849$, partial $\eta2 = .151$. Follow-up univariate ANOVAs indicated that positive emotions, negative emotions, and classroom engagement were significantly different for the two classrooms, $F(1, 99) = 7.183$, $p = .009 < .05$, partial $\eta2 = .068$, $F(1, 99) = 6.369$, $p = .013 < .05$, partial $\eta2 = .060$, and $F(1, 99) = 17.004$, $p < .001$, partial $\eta2 = .147$, respectively. There were no significant differences in chemistry self-efficacy beliefs, $F(1, 99) = 3.382$, $p = .069 > .05$, partial $\eta2 = .033$. These findings suggest that the computer simulation might positively influence students in the two classrooms to different extents even though pedagogical strategies were similar. One possible explanation is that the computer simulation influenced students' affective perceptions both at the individual level and at the classroom level. On one hand, the incitement of positive emotions (e.g., curiosity) might attract the individual's attention to dynamic animations in the computer simulation. On the other hand, students' positive emotions might create an overexciting atmosphere and decrease the classroom order. At the individual level, one student mentioned that the over-excitement decreased classroom engagement, which supports the previous argument that students' positive emotions may reduce cognitive resources available and distract attention away from academic tasks (Meinhardt & Pekrun, 2003). In other words, the excitement may lead to shallow processing of information and reduce the motivation for deep involvement. At the classroom level, where students interact and influence each other, the influence of positive emotions on engagement is more complicated. The survey data and classroom observation showed that the demonstration of dynamic animations increased positive emotions, which also influenced others' behaviors and the whole instructional or learning environments (Pekrun et al., 2002). The relatively relaxed classroom atmosphere potentially decreased classroom order. In the interview, both teachers mentioned the problem of "controlling the scene". Contextual factors such as the noise in the classroom might distract students at the back of classrooms, who struggled to hear the authority figure - teacher's hints and directions, from learning the content. As classroom management is one important issue in large classrooms, one tentative inference is that the computer simulation's function of increasing chemistry self-efficacy beliefs is more appealing than the effect of inciting positive emotions in large Chinese classrooms. Teacher guidance plays a critical role in transforming behavioral engagement into cognitive engagement.

Conclusions, Limitations, and Implications

Previous literature has documented two pedagogical strategies for integrating computer simulations: An inquiry tool for developing scientific practices and a visualization tool for promoting content understanding. Even though using computer simulations as an inquiry tool can maximize advantages of technology, such pedagogical strategy is susceptible to classroom contexts such as the number of students, class time, technology equipment, and teachers' pedagogical goals. This study narrated how the two Chinese chemistry teachers integrated the computer simulations as a visualization tool and the impact on students' various affective perceptions in real classroom contexts. Therefore, it contributes to educational research in the following two ways: First, it provides one pedagogical strategy for integrating computer simulations into typical highly structured and large Chinese classrooms. Second, it shows that this specific way of integrating computer simulations can potentially deepen conceptual understanding and improve specific affective perceptions. The survey data showed that computer simulations significantly and positively increased students' chemistry self-efficacy beliefs and positive emotions. However, there were no significant differences in students' negative emotions and classroom engagement.

Based on the findings of this study, I recommend that computer simulations can be and should be integrated into large Chinese classes to support traditional chemistry instruction. The dynamic and concrete format of representing abstract scientific concepts can facilitate the cognitive process of constructing mental models and benefit students who are disadvantaged or disengaged in lecture-dominated classrooms. Even though the interview data revealed that dynamic animations might also distract students from learning science content, such concerns can be reduced or avoided if teachers can give effective and timely directions, which guide students to connect simulation features with content knowledge. Teachers' pedagogical strategies are critical for managing the tradeoff between possible advantages and disadvantages of computer simulations. Since this study did not interfere with teachers' instructional decisions, the two Chinese teachers integrated the computer simulation as a visualization tool to review content knowledge. Despite various contextual challenges embedded within the Chinese education system (e.g., rigid curriculum standards and heavy curriculum tasks), it is still possible to reform the traditional lecture-based instruction and integrate computer simulations as an interactive demonstration tool to support inquiry practices. In the future, researchers can compare how different pedagogical strategies of integrating computer simulations influence students' affective perceptions in large Chinese classrooms. The current study also encourages further research to explore what kinds of pedagogical practices of using computer simulations are more effective for engaging students in science classes. In summary, this study is worthwhile for those who are interested in utilizing computer simulations to create emotionally pleasant classroom climates and in improving the effectiveness of integrating computer simulations in different classroom contexts.

Interpretations of the results should take the following limitations into consideration. First, this study was exploratory in nature where the computer simulation was integrated for only one chemistry class. The short period of intervention could not accurately illuminate the longitudinal effect of integrating computer simulations on students' affective perceptions. Specifically, it is unknown whether the novelty derived from the technology itself is meaningful or not in the long term. Second, due to the availability of identification data, I used the independent sample t tests to compare means of four affective variables between pre- and post-surveys. The same group of participants violated the assumption of independence, which might lead to the failure of detecting the difference that was significant in dependent t tests. Third, the interview data indicated that students held different goal orientations when reporting the change of self-efficacy beliefs. Due to the lack of identification information, this study could not control students' goal orientations before comparing their chemistry self-efficacy beliefs. Based on thesee limitations, I provide two suggestions for future research directions. First, the larger and longitudinal data should be collected to systematically examine the influence of computer simulations on students' affective perceptions and academic performance. Second, researchers can first code students' goal orientations into two categories (i.e., learning versus performance) and then examine the significance of difference in chemistry self-efficacy beliefs between the two groups.

Appendices: Interview Questions and Open-Response Questions

Appendix 1: Interview Questions (English)

1. Please tell me a little bit about your general chemistry class.
2. How do you feel about your general chemistry class?
3. Please tell me a little bit about your simulation-based chemistry class.
4. How do you feel about your simulation-based chemistry class?
5. What is the influence of using computer simulations in the classroom? (e.g., interest in course materials, involvement in the lecture, interaction with other students, achieving course objectives, participation in classroom discussions, teachers' responses to concepts that might not have been understood, engagement and involvement)
6. What do you think of using computer simulations in the long term?

Appendix 2: Open-Response Questions (English)

1. Please describe your feelings about simulation-based chemistry class in general.
2. Please check the following words that can accurately describe your feelings in simulation-based chemistry class.
    ☐ Happy
    ☐ Excited
    ☐ Nervous
    ☐ Anxious
    ☐ Proud
    ☐ Shamed
    ☐ Annoyed
    ☐ Angry
    ☐ Enjoyable
    ☐ Other
3. Please describe the reason for above feelings.

4. What is the influence of using computer simulations on your feelings in chemistry class?
5. What is the influence of using computer simulations on your confidence of learning chemistry?
6. What is the influence of using computer simulations on your engagement in chemistry class?

References

Bandura, A. (1997). *Self-efficacy: The exercise of control*. New York: Freeman.

Bell, B. S., & Kozlowski, W. J. (2002). Goal orientation and ability: Interactive effects on self-efficacy, performance, and knowledge. *Journal of Applied Psychology, 87*(3), 497.

Bell, T., Urhahne, D., Schanze, S., & Ploetzner, R. (2010). Collaborative inquiry learning: Models, tools, and challenges. *International Journal of Science Education, 32*(3), 349–377.

Bong, M., & Skaalvik, E. M. (2003). Academic self-concept and self-efficacy: How different are they really? *Educational Psychology Review, 15*(1), 1–40.

Clarke, J., & Dede, C. (2005, April). *Making learning meaningful: An exploratory study of using multi-user environments (MUVEs) in middle school science*. In *American Educational Research Association Conference*, Montreal, Canada.

Cohen, J. (1988). *Statistical power analysis for the behavioral sciences*. Hillsdale, NJ: Earlbaum.

Corbin, J., & Strauss, A. (1990). Grounded theory research: Procedures, canons and evaluative criteria. *Zeitschrift für Soziologie, 19*(6), 418–427.

Dalgety, J., Coll, R. K., & Jones, A. (2003). Development of chemistry attitudes and experiences questionnaire (CAEQ). *Journal of Research in Science Teaching, 40*(7), 649–668.

Dweck, C. S. (1986). Motivational processes affecting learning. *American Psychologist, 41*(10), 1040–1048.

Finkelstein, N., Adams, W., Keller, C., Perkins, K., & Wieman, C. (2006). High-tech tools for teaching physics: The physics education technology project. *Journal of Online Learning and Teaching, 2*(3), 110–120.

Fredricks, J. A., Blumenfeld, P. C., & Paris, A. H. (2004). School engagement: Potential of the concept, state of the evidence. *Review of Educational Research, 74*(1), 59–109.

Gist, M. E., & Mitchell, T. R. (1992). Self-efficacy: A theoretical analysis of its determinants and malleability. *Academy of Management Review, 17*(2), 183–211.

Gong, X., & Bergey, B. W. (2020). The dimensions and functions of students' achievement emotions in Chinese chemistry classrooms. *International Journal of Science Education, 42*(5), 835–856.

Gong, X., Wei, B., Bergey, B. W., & Shockley, E.T. (2023). Unpacking chemistry teachers' pedagogical reasoning and decisions for a PHET simulation: A TPACK perspective. *Journal of Chemical Education, 100*(1), 34–44.

Hennessy, S., Deaney, R., & Ruthven, K. (2006). Situated expertise in integrating use of multimedia simulation into secondary science teaching. *International Journal of Science Education, 28*(7), 701–732.

Hennessy, S., Wishart, J., Whitelock, D., Deaney, R., Brawn, R., La Velle, L., ... Winterbottom, M. (2007). Pedagogical approaches for technology-integrated science teaching. *Computers & Education, 48*(1), 137–152.

Hsu, Y. S., & Thomas, R. A. (2002). The impacts of a web-aided instructional simulation on science learning. *International Journal of Science Education, 24*(9), 955–979.

Jimoyiannis, A., & Komis, V. (2001). Computer simulations in physics teaching and learning: A case study on students' understanding of trajectory motion. *Computers & Education, 36*(2), 183–204.

Ketelhut, D. J. (2007). The impact of student self-efficacy on scientific inquiry skills: An exploratory investigation in River City, a multi-user virtual environment. *Journal of Science Education and Technology, 16*(1), 99–111.

Ketelhut, D. J., & Nelson, B. C. (2010). Designing for real-world scientific inquiry in virtual environments. *Educational Research, 52*(2), 151–167.

Kiboss, J. K., Ndirangu, M., & Wekesa, E. W. (2004). Effectiveness of a computer-mediated simulations program in school biology on pupils' learning outcomes in cell theory. *Journal of Science Education and Technology, 13*(2), 207–213.

Laurillard, D. (1992). Learning through collaborative computer simulations. *British Journal of Educational Technology, 23*(3), 164–171.

Manfra, M. M., & Hammond, T. C. (2008). Teachers' instructional choices with student-created digital documentaries: Case studies. *Journal of Research on Technology in Education, 41*(2), 223–245.

McKagan, S. B., Perkins, K. K., Dubson, M., Malley, C., Reid, S., LeMaster, R., & Wieman, C. E. (2008). Developing and researching PhET simulations for teaching quantum mechanics. *American Journal of Physics, 76*(4), 406–417.

Meinhardt, J., & Pekrun, R. (2003). Attentional resource allocation to emotional events: An ERP study. *Cognition and Emotion, 17*, 477–500.

Meluso, A., Zheng, M., Spires, H. A., & Lester, J. (2012). Enhancing 5th graders' science content knowledge and self-efficacy through game-based learning. *Computers & Education, 59*(2), 497–504.

Pekrun, R., Goetz, T., & Frenzel, A. C. (2005). *Academic Emotions Questionnaire – Mathematics (AEQ-M): User's manual.* Munich, Germany: University of Munich, Department of Psychology.

Pekrun, R., Goetz, T., Titz, W, & Perry, R. P. (2002). Positive emotions in education. In E. Frydenberg (Ed.), *Beyond coping: Meeting goals, visions, and challenges* (pp. 149–174). Oxford, UK: Elsevier.

Phillips, J. M., & Gully, S. M. (1997). Role of goal orientation, ability, need for achievement, and locus of control in the self-efficacy and goal-setting process. *Journal of Applied Psychology, 82*(5), 792–802.

Rutten, N., van der Veen, J. T., & van Joolingen, W. R. (2015). Inquiry-based whole-class teaching with computer simulations in physics. *International Journal of Science Education, 37*(8), 1225–1245.

Skinner, E., Furrer, C., Marchand, G., & Kindermann, T. (2008). Engagement and disaffection in the classroom: part of a larger motivational dynamic? *Journal of Educational Psychology, 100*(4), 765.

Smetana, L. K., & Bell, R. L. (2012). Computer simulations to support science instruction and learning: A critical review of the literature. *International Journal of Science Education, 34*(9), 1337–1370.

Stern, L., Barnea, N., & Shauli, S. (2008). The effect of a computerized simulation on middle school students' understanding of the kinetic molecular theory. *Journal of Science Education and Technology, 17*(4), 305–315.

Van der Meij, J., & de Jong, T. (2006). Supporting students' learning with multiple representations in a dynamic simulation-based learning environment. *Learning and Instruction, 16*(3), 199–212.

Wells, J. D., Campbell, D. E., Valacich, J. S., & Featherman, M. (2010). The effect of perceived novelty on the adoption of information technology innovations: A risk/reward perspective. *Decision Sciences, 41*(4), 813–843.

Wigfield, A., & Cambria, J. (2010). Students' achievement values, goal orientations, and interest: Definitions, development, and relations to achievement outcomes. *Developmental Review, 30*(1), 1–35.

Yin, R. K. (2003). *Case study research: design and methods* (3rd ed.). Thousands Oaks, CA: Sage.

5 University Science-Major Students' Achievement Emotions in Emergency Remote Learning during COVID-19 Pandemic

A Latent Profile Analysis

Introduction

The COVID-19 pandemic has transformed school systems from face-to-face instruction to online instruction due to school closures. The change in educational environments greatly influences higher education students' academic performance and mental health worldwide (Kecojevic, Basch, Sullivan, & Davi, 2020; Lee, 2020). Social isolation resulting from the COVID-19 pandemic and students' interests or difficulties in adapting to online learning environments may positively or negatively impact students' emotions related to learning (e.g., Horesh & Brown, 2020). What is more, university students who vary in their abilities or strategies of emotional regulation may handle the transformation in different ways, leading to different learning outcomes. Yet, previous studies have adopted variable-centered approaches to examine how various types of achievement emotions predict academic learning outcomes independently and neglect possible interactions among positive and negative emotions (e.g., Artino & Jones, 2012; Heckel & Ringeisen, 2019). As such, it is worthwhile to adopt person-centered approaches to identify university students' emotional profiles in emergency remote learning during the COVID-19 pandemic and investigate how profile memberships relate to students' satisfaction and emotion regulation strategies. The results of this study may provide implications for the design of online learning activities and pedagogical support of emotional regulation in online settings.

Literature Review

Control-Value Theory and Achievement Emotions in Online Settings

As the theoretical framework of this study, the control-value theory postulates that features of environments related to individuals' control and value appraisals influence students' perceptions of achievement emotions and performance in academic settings (Pekrun, 2006). These features include the quality of classroom instruction, autonomy support versus control, students' perceived achievement expectations, feedback and consequences of achievement, and

DOI: 10.4324/9781032649665-8

social relatedness or support in academic-related interactions (Pekrun, 2000; Pekrun, Goetz, Titz, & Perry, 2002). Compared with regular face-to-face classroom instruction, online instruction owns its unique characteristics and creates more complicated or unstructured learning environments. It is regarded as a double-edged sword that provides opportunities and challenges for student learning (Lee & Choi, 2011). For example, the quality of online instruction is influenced by teachers' willingness and expertise to use online resources and platforms, the stability of network connections, and the appropriateness of pedagogical strategies for organizing instructional activities and monitoring student progress. As Chinese teachers have not been systematically trained for adopting fully online instruction, they may encounter challenges in selecting appropriate technology tools and technical problems since each technology tool has its affordances and limitations. What is more, various distractions in online learning environments, such as social media, video games, and instant messages, aggravate the problem of cognitive engagement (Gaudreau, Miranda, & Gareau, 2014; Moore & Kearsley, 2012), specifically for students who lack self-regulatory abilities. The absence of face-to-face communication influences how the individual student regulates his or her learning behaviors and interactions with teachers or peers. Simultaneously, as online learning is more flexible and convenient than traditional fixed classroom locations, students may feel less pressure due to reduced peer competition and teacher supervision. What is more, rich online learning resources such as Massive Open Online Courses (MOOCs) or other supplementary materials in electronic format provide students the flexibility or autonomy to develop personalized approaches to academic learning. In summary, promises and challenges embedded in online environments may contribute to eliciting various positive or negative emotional responses.

The conceptualization of achievement emotions in control-value theory and its antecedents and outcomes have been systematically described in previous chapters. Based on existing theoretical hypotheses, survey data was collected in Chapter 2 to examine Chinese high school students' achievement emotions in regular science classrooms, and results highlighted the role of cultural values in shaping students' emotional perceptions. In order to avoid repetition and save space, I will focus on recent studies that have examined students' emotions in online settings in this section. At present, most of research on achievement emotions pays attention to face-to-face classroom settings, while few studies have examined students' achievement emotions in online settings (Buhr, Daniels, & Goegan, 2019; Heckel & Ringeisen, 2019). Similar to face-to-face classroom instruction, students experience a series of positive and negative emotions in online environments including enjoyment, stress, tiredness, frustration, confusion, anxiety, boredom, fear, and isolation (e.g., Buhr et al., 2019; Parker et al., 2021; Racanello et al., 2022).

However, differences in the intensity or frequency of experienced emotions between face-to-face and online settings may depend on specific types of emotions under consideration. Stephan, Markus, and Gläser-Zikuda (2019) found that the online group reported higher boredom, anxiety, and anger but less enjoyment than the on-campus group while no significant differences were found for hope, pride, shame, and hopelessness. In accordance with theoretical assumptions, recent studies have provided empirical evidence that students' perceived achievement emotions mediate the relationship among control or value appraisals, learning behaviors and academic outcomes in online settings. For example, Heckel and Ringeisen (2019) reported that German university students' perceived pride positively predicted satisfaction while anxiety negatively predicted satisfaction in online learning environments. Artino and Jones (2012) examined the relationship between three discrete emotions (e.g., enjoyment, boredom, and frustration) and two types of self-regulated learning behaviors (e.g., elaboration and metacognition) in an online course. Results showed that American undergraduate students' perceived enjoyment was a positive predictor of elaboration and metacognition. In contrast, the emotion of frustration only positively predicted metacognition while boredom negatively predicted metacognition. Parker et al. (2021) used latent profile analysis to identify students' multifaceted motivational profiles in an online university course. They reported that three motivation profiles emerged: High control-enjoyment, low control-boredom, and low value-boredom. The profile membership further predicted students' achievement perceptions (i.e., perceived success and expected grades) and academic performance.

In Eastern cultural contexts, You and Kang (2014) found that Korean college students' perceived enjoyment was a positive and significant predictor of self-regulated learning while its relations with anxiety and boredom were not significant in online learning. Chinese university students' negative emotions negatively and significantly predicted online learning engagement while positive emotions positively predicted online learning engagement (Deng, Lei, Guo, Li, Ge, & Hu, 2022). Ding and Zhao (2020) further divided engagement in MOOCs into two types: Engagement with videos and engagement with assignments. Results showed that Chinese university students' enjoyment, excitement, boredom, and annoyance significantly predicted video engagement while only excitement and annoyance significantly predicted assignment engagement. These findings across different cultural contexts suggest that underlying mechanisms of how positive and negative emotions influence students' learning behaviors in online settings are complex and worth further investigation.

Emotional Regulation Strategies and Cultural Values

Students' abilities to regulate emotions play an important role in predicting their mental health and academic performance during the COVID-19 pandemic. Emotional regulation occurs when the individual activates a goal to modify the generative process of emotional responses (Gross, 2013; Gross, Sheppes, & Urry, 2011). Gross and John (2003) described two types of emotional regulation strategies: Cognitive reappraisal and expression suppression.

Cognitive reappraisal refers to the cognitive change of constructing an emotion-eliciting situation in the way that modifies its potential emotional impact (Lazarus & Alfert, 1964). It is considered as an antecedent-focused strategy that often occurs before the generation of emotional responses. For example, students who feel big pressure to give presentations in the classroom may tell themselves that it is an opportunity to improve their communication abilities. Such autosuggestion may reduce students' perceptions of anxiety and pressure. Expressive suppression is a response-focused strategy that inhibits outward emotional or behavioral expressions (Gross, 2014). Compared with cognitive reappraisal, it is used to regulate emotions after the generation of emotions. For example, students in Eastern contexts may hide or constrain activating positive or negative emotions to keep calm in regular classroom settings. The preceding two strategies of regulating emotions exert influence on individuals' well-being and life satisfaction in different ways (Dryman & Heimberg, 2018; Gross & John, 2003).

Whereas the strategy of cognitive reappraisal is thought to have a healthier profile of affective, cognitive, and social consequences than the strategy of expressive suppression (John & Gross, 2004), the preceding two emotional regulation strategies are not unrelated or isolated. Instead, the individual may employ two approaches simultaneously when facing specific tasks. For example, Lee et al. (2016) examined relations among teachers' discrete emotions, emotion regulation strategies, and emotional labor strategies. Results showed that the suppression strategy and the reappraisal strategy were positively correlated. The reappraisal strategy was positively related to enjoyment while the suppression strategy was positively related to anxiety. By contrast, neither of the two strategies was significantly related to pride, anger, and frustration. However, the empirical evidence on the relationship between achievement emotions and emotional regulation strategies is inconsistent. Sorić, Penezić, and Burić (2013) reported that both the cognitive reappraisal strategy and the expressive suppression strategy positively predicted students' perceived negative emotions (e.g., unhappiness and humiliation) with the exception of anger. Gross and John (2003) found that students who used the cognitive reappraisal strategy experienced or expressed greater positive emotions and less negative emotions. In contrast, students who frequently used the expression suppression strategy experienced or expressed lower positive emotions but greater negative emotions. Despite these inconsistent findings, it is widely assumed that the cognitive reappraisal strategy has the potential to promote students' academic achievement and engagement (John & Gross, 2004). As such, it is healthier for individuals to regulate emotions with the strategy of cognitive reappraisal rather than the strategy of expression suppression.

Besides individual characteristics, students' tendencies or social consequences of using different emotional regulation strategies also depend on specific values or norms defined in different cultural contexts. Context plays a crucial role in influencing students' experience or expression of emotions and the form, frequency, and function of emotional regulation strategies

(Aldao, 2013; Soto, Perez, Kim, Lee, & Minnick, 2011). As described in the first chapter, whereas Western cultural values encourage direct and open expressions of emotions (Oyserman, Coon, & Kemmelmeier, 2002), Eastern cultural values encourage students to express low-activating emotions (e.g., calmness) and constrain certain high-activating emotions (e.g., anger) (Butler & Gross, 2009; Butler, Lee, & Gross, 2007, 2009). In Western cultures, students tend to maximize desirable positive emotions and minimize undesirable negative emotions (Eid & Diener, 2001; Kitayama, Markus, & Kurokawa, 2000). In Eastern cultures, those culturally desirable emotions are not equivalent to positive emotions, which raises the issue of the nature of the relationship between positive emotions and negative emotions (i.e., independence versus bipolarity) (Barrett & Russell, 1998). Eastern students tend to find a middle way and experience a dialectical balance between positive emotions and negative emotions (Miyamoto & Ma, 2011). These cross-cultural differences in emotional experience and expression are supported with rich empirical evidence. Kitayama et al. (2000) found that Western students' frequency of reporting positive emotions was much higher than negative emotions, but such difference disappeared among Eastern students. By contrast, Eastern students tend to experience both positive and negative emotions moderately frequently (Miyamoto & Ryff, 2011).

What is more, students' emotional experience or expression are the consequence of applying different emotional regulation strategies. At the individual level, students' selections of emotional regulation strategies may vary across age and older students are more likely to use the cognitive reappraisal strategy. Liu and Sang (2020) found that middle school students reported to use the expressive suppression strategy more frequently than high school students when regulating positive and negative emotions. Positive emotions positively and significantly predicted the use of the expressive suppression strategy while the effect of negative emotions on the expressive suppression strategy was not significant. At the cultural level, cultural values moderate the relationship between specific emotional regulation strategies (e.g., expressive suppression) and mental health (Hu et al., 2014). Soto et al. (2011) found that Eastern students reported using suppression significantly more frequently than Western peers. While the expressive suppression strategy resulted in depressed mood and life satisfaction for Western students, such links disappeared for Eastern students. These findings provide theoretical and empirical basis for examining Chinese students' achievement emotions and their relations with emotional regulation strategies and course satisfaction, especially in online learning environments.

The Current Study

At present, little attention has been paid to students' achievement emotions and emotional regulation strategies in online academic settings, especially in Eastern contexts (Buhr et al., 2019; Liu & Sang, 2020). What is more, the association between achievement emotions and emotional regulation strategies warrants further clarification (Gross, 2015; Gross & Jazaieri, 2014). As the variable-centered approach showed discrepant findings regarding the relationship between emotions and regulation strategies, in this study, a person-centered approach involving latent profile analysis was adopted to investigate how various achievement emotions with different valences (i.e., enjoyment, pride, anxiety, and anger) formed different patterns in online learning environments and how profile memberships predicted students' satisfaction with online learning and emotional regulation strategies. Based on the classification of achievement emotions in control-value theory, enjoyment, pride, anxiety, and anger are a positive activity emotion, a positive outcome emotion, a negative outcome emotion, and a negative activity or outcome emotion, respectively.

Methods

Participants

Participants (N = 356) were recruited from a large public university in northern China. They all majored in science domains (e.g., physics, biology, and chemistry) and enrolled in the same teacher education course in the fall semester of the 2020–2021 academic year. Of the university students, 14.9% were in the sophomore year and 85.1% were in the junior year. The average age was 20 years old (SD = 10 months). Among them, 82.6% were self-identified as females and 9.6% were identified as minorities. Students completed either the web-based or paper-based survey voluntarily during the teacher education course. The average time to complete the survey was 8 minutes.

Measures

The survey measured three variables including achievement emotions, online learning satisfaction, and emotional regulation strategies with Likert scale items. Students selected their options on a 6-point Likert scale with anchors at 1 (strongly disagree) and 6 (strongly agree).

Achievement Emotions

Students' achievement emotions were assessed using the Chinese version of the Achievement Emotion Questionnaire (AEQ) developed by Pekrun, Goetz, & Frenzel (2005). The instrument's reliability and validity have been demonstrated in Eastern and Western countries. In this study, survey items were modified by changing "math" to "online classes". Only items measuring students' four types of emotions (i.e., enjoyment, pride, anxiety, and anger) were included in the final analysis. Table 5.1 shows internal reliabilities and example items of four subscales used in the following analysis. Before the latent profile analysis, standardized mean scores for four types of achievement emotions were calculated.

Table 5.1 Scales and example items

Scale Sets	Construct	Alpha (No. of Items)	Example Item
Set A: Online Achievement Emotions	Enjoy	0.723(5)	I enjoy my online class so much that I am strongly motivated to participate.
	Pride	0.811(4)	I think I can be proud of my knowledge learned in online classes.
	Anxiety	0.798(7)	I worry that the content in online classes is much too difficult for me.
	Anger	0.685(4)	My online-class homework makes me angry.
Set B: Emotion Regulation	Suppression	0.789(3)	I control my emotions by not expressing them in online classes.
	Reappraisal	0.756(3)	When I want to feel less anxiety in online classes, I change the focus of attention.
Set C: Online Class Satisfaction	Satisfaction	0.838(3)	I feel that online classes serve my needs well.
	Dissatisfaction	0.787(3)	Conducting the course via the Internet makes it more difficult.

Satisfaction and Dissatisfaction

Students' perceived satisfaction with online classes was assessed with six items developed by Arbaugh (2000). The research team translated the scale to Chinese and then one education researcher who was fluent in English back-translated the Chinese version to English, and the research team verified the match with the English version. In order to examine the construct structure, I conducted exploratory principal axis factoring with an oblimin rotation with SPSS. Eigenvalues, scree plot, percent variance explained by each factor, and factor loadings in the pattern matrix were used to evaluate the solution. Results showed that two factors explained a total of 60.867% of the variance, with the first factor explaining 48.832% of the variance. The first factor included three items and measured students' perceived satisfaction and tendency to choose online instruction in the future while the second factor included three items and measured students' perceived disappointment and repulsion of making the same decision. Therefore, I labeled the first factor as *online-class satisfaction* and the second factor as *online-class dissatisfaction*. The two subscales demonstrated good reliabilities (see Table 5.1 for Cronbach's alphas, number of items, and example items).

Emotional Regulation Strategy

Students' emotional regulation strategies were measured by six items adapted from the Emotion Regulation Questionnaire (ERQ) developed by Gross and

John (2003). Three items measured the cognitive reappraisal strategy while three items measured the expressive suppression strategy. I used the same language translation process described for the satisfaction scale. Survey items were modified to apply to online class settings and to match the typical Chinese expression. For example, "I control my emotions by changing the way I think about the situation I'm in" was revised as "I regulate my emotions by changing the focus of attention". Cronbach's alphas, number of items, and example items are listed in Table 5.1.

Analytic Plan

First, the eight variables were screened for normality and the presence of outliers. Then latent profile analysis (LPA) was conducted to examine students' emotional profiles that emerged from the data. LPA is a person-centered statistical procedure used to identify subgroups of individuals within populations that share similar characteristics (Hagenaars & McCutcheon, 2002). The underlying assumption is that profile membership can explain patterns of individuals' scores on indicator variables in the scale. In this study, I used Mplus and tested a series of sequential models (e.g., 1, 2, 3, 4, and 5 latent profiles). In order to determine the best solution, I examined a set of fit statistics including the Entropy index, Akaike Information Criterion (AIC), Bayesian Information Criterion (BIC), sample-size Adjusted Bayesian Information Criterion (ABIC), and adjusted Lo-Mendel-Rubin adjusted Likelihood Ratio Test (LMR-LRT). Entropy values approaching 1.0 indicate a clearer separation of participants into profiles (Nylund-Gibson, Grimm, Quirk, & Furlong, 2014). Lower values of the AIC, BIC, and adjusted BIC tests represent a better-fit model (Schwarz, 1978). Significant values generated by the LMR-LRT indicate that the tested model with more latent profiles has a significantly better fit than a model with fewer profiles (Lo, Mendell, & Rubin, 2001). In addition to information criterion statistics, I also evaluated interpretability and parsimony to guarantee practical and theoretical usefulness of latent profiles (Hickendorff, Edelsbrunner, McMullen, Schneider, & Trezise, 2018). After identifying the number of latent profiles, I used a one-way multivariate analysis of variance (MANOVA) to investigate how profile membership related to students' perceived satisfaction and emotional regulation strategies. Partial eta squared (η_p^2) was reported to indicate the proportion of total variance in a dependent variable that is explained by profile membership after partialing out effects of other variables (Richardson, 2011).

Results

Descriptive statistics and bivariate correlations of measured variables are listed in Tables 5.2 and 5.3, respectively. As shown in Table 5.2, the values of kurtosis and skewness were less than 1.5, which indicated the normal distribution for all variables (Kline, 1998). Average scores ranged from 3.10 (i.e., anger) to 4.16 (i.e., pride). As showed in Table 5.3, four types of emotions (i.e., enjoyment,

Table 5.2 Descriptive statistics of variables

Variables	M	SD	Skewness	Kurtosis
1. Enjoyment	3.57	0.79	0.09	0.27
2. Pride	4.16	0.86	−0.34	0.18
3. Anxiety	3.78	0.83	−0.15	0.11
4. Anger	3.10	0.88	0.16	0.41
5. Reappraisal	4.10	0.89	−0.16	0.58
6. Suppression	3.67	1.00	−0.24	0.30
7. Satisfaction	3.64	1.08	0.04	−0.07
8. Dissatisfaction	3.49	1.13	−0.02	−0.50

Table 5.3 Bivariate correlations among variables

Variables	1	2	3	4	5	6	7
1. Enjoyment	–						
2. Pride	.629**	–					
3. Anxiety	.076	.228**	–				
4. Anger	.035	.141**	.521**	–			
5. Reappraisal	.014	.177**	.256**	.218**	–		
6. Suppression	.319**	.266**	.170**	.136*	.184**	–	
7. Satisfaction	.676**	.463**	−.117*	−.118*	−.002	.199**	–
8. Dissatisfaction	−.269**	−.075	.469**	.406**	.270**	.065	−.510**

* *p* < 0.05.
** *p* < 0.01.

pride, anxiety, and anger) were positively correlated with each other and correlation coefficients varied between 0.035 and 0.629. Except for the correlation between enjoyment and the cognitive reappraisal strategy (r = .014), four types of emotions were positively and significantly correlated with the two types of emotional regulation strategies and correlation coefficients varied between 0.136 and 0.319. What is more, students' perceived satisfaction with online learning was positively correlated with positive emotions and the expressive suppression strategy while negatively correlated with negative emotions. In contrast, students' dissatisfaction with online learning was positively correlated with negative emotions and the cognitive reappraisal strategy while negatively correlated with the feeling of enjoyment.

Latent Profile Analysis and Model Selection (RQ1)

The latent profile analysis was used to identify university students' emotional profiles. The fit indices, entropy statistics, and LMR-LRT *p* value were used to determine the number of latent profiles. Table 5.4 showed that values for AIC, BIC, and ABIC decreased as the number of profiles increased. The LMR-LTR tests were significant for the 2-profile solution and 4-profile solution.

Table 5.4 Indices for profile selection

Model	AIC	BIC	Adjusted BIC	Entropy	LMR-LTR p value
1 Profile	4030.428	4061.428	4036.048	–	–
2 Profile	3923.626	3974.001	3932.759	0.721	0.0473*
3 Profile	3862.559	3932.308	3875.204	0.745	0.0972
4 Profile	**3798.068**	**3887.191**	**3814.225**	**0.761**	**0.0112***
5 Profile	3768.474	3876.972	3788.144	0.795	0.0983

Note: Bold font indicates the selected best fitting model.
* p-value significant at the 0.05 level.

After considering the above criteria and theoretical usefulness, I selected the 4-profile solution with the entropy value of 0.76. The model suggested four unique and theoretically meaningful latent profiles. Table 5.5 and Figure 5.1 illustrate each latent profile's standardized mean scores and standard errors for perceived enjoyment, pride, anxiety, and anger. For interpretation purposes,

Table 5.5 Standardized means and standard errors by latent profiles

Model	Profiles (Percent of total sample)							
	1 (57.9%)		2 (26.4%)		3 (4.2%)		4 (11.5%)	
	M	SE	M	SE	M	SE	M	SE
Enjoy	−0.251	0.106	0.734	0.138	1.562	0.443	−1.112	0.143
Pride	−0.086	0.108	0.647	0.112	1.211	0.192	−1.523	0.161
Anxiety	0.357	0.075	−0.544	0.174	1.484	0.323	−1.048	0.209
Anger	0.382	0.09	−0.662	0.153	1.474	0.373	−0.904	0.175

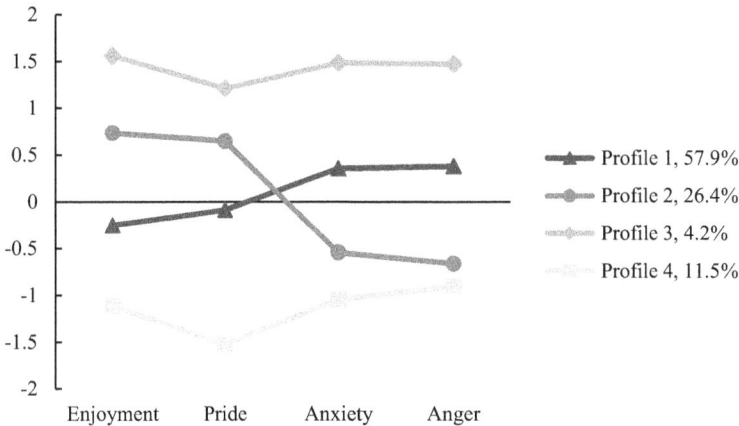

Figure 5.1 Standardized means by latent profiles.

profile characteristics were described and labeled based on variable means relative to the whole sample and original response scales (Wormington & Linnenbrink-Garcia, 2017). In original response scales, variables were assessed with a 6-point scale. Students' scores were characterized as follows: Low (1.0–1.9), moderately low (2.0–2.9), moderate (3.0–3.9), moderately high (4.0–4.9), high (5.0–6.0). The magnitudes of standardized scores were characterized as follows: Low (below −1 SD), moderately low (−1 SD to −0.5SD), moderate (−0.49SD to +0.49 SD), moderately high (0.5SD to 1 SD), high (above 1 SD).

Profile 1: Moderate Positive Emotions, Moderate Negative Emotions

The first and largest profile, comprising 57.9% of the sample, was the most common profile in the sample. It was characterized by moderate enjoyment ($M = 3.37$) and moderate pride ($M = 4.07$). Both enjoyment and pride were slightly below sample average scores. This profile was also characterized by moderately high anxiety ($M = 4.12$) and moderate anger ($M = 3.46$), which were slightly above sample average scores. Average scores of both positive emotions and negative emotions were quite close to the overall means of the sample (−0.251 to 0.382).

Profile 2: Moderately High Positive Emotions, Moderately Low Negative Emotions

The second profile, comprising 26.4% of the sample, was characterized by moderately high enjoyment ($M = 4.19$) and moderately high pride ($M = 4.77$). Both enjoyment and pride were approximately half standard deviation above sample average scores. This profile was also characterized by moderate anxiety ($M = 3.23$) and moderately low anger ($M = 2.44$), which were approximately half standard deviation below sample average scores (−0.662 to 0.734).

Profile 3: High Positive Emotions, High Negative Emotions

The third profile, comprising 4.2% of the sample, was the smallest profile in the sample. It was characterized by moderately high enjoyment ($M = 4.91$), high pride ($M = 5.25$), high anxiety ($M = 5.13$), and moderately high anger ($M = 4.50$). Average scores of both positive emotions and negative emotions were the highest among all latent profiles, with more than one standard deviation above sample average scores (1.211 to 1.562).

Profile 4: Low Positive Emotions, Low Negative Emotions

The fourth profile, comprising 11.5% of the sample, was characterized by moderately low enjoyment ($M = 2.65$), moderately low pride ($M = 2.80$), moderately low anxiety ($M = 2.84$), and moderately low anger ($M = 2.26$). Average scores of both positive emotions and negative emotions were the lowest among all latent profiles, approximately one standard deviation below sample average scores (−0.904 to −1.523).

Relations between Profiles, Online-Class Satisfaction, and Emotional Regulation Strategies

In this section, I examined how latent profile memberships related to students' perceived satisfaction, dissatisfaction, and strategies of emotional regulation in online learning environments. Before comparing group differences of these variables among four latent profiles, the Box's Test of Equality of Covariance Matrices showed that there were significant differences between the covariance matrices ($M = 74.72$, F [30, 10233.327] = 2.353, $p < .001$). Therefore, *Pillai's Trace* tests were used in the following MANOVA output. Table 5.6 and Figure 5.2 show estimated marginal means and standard errors for four latent groups' outcome variables. Levene's tests were not statistically significant for online satisfaction (F [3, 346] = 1.035, $p = .706$) and dissatisfaction (F [3, 346] = 1.035, $p = .377$). In contrast, Levene's tests were statistically significant for the cognitive reappraisal strategy (F [3, 346] = 4.035, $p = .005$) and

Table 5.6 Estimated marginal means and standard errors for dissatisfaction, satisfaction, reappraisal, and suppression by latent profile

Model	Profiles (Percent of the Total Sample)							
	1 (57.9%)		2 (26.4%)		3 (4.2%)		4 (11.5%)	
	M	SE	M	SE	M	SE	M	SE
Dissatisfaction	3.799	.073	2.761	.108	4.244	.267	3.358	.162
Satisfaction	3.381	.066	4.391	.098	4.556	.241	2.846	.146
Reappraisal	4.160	.062	3.986	.091	4.800	.226	3.805	.137
Suppression	3.630	.069	3.725	.103	4.600	.254	3.325	.154

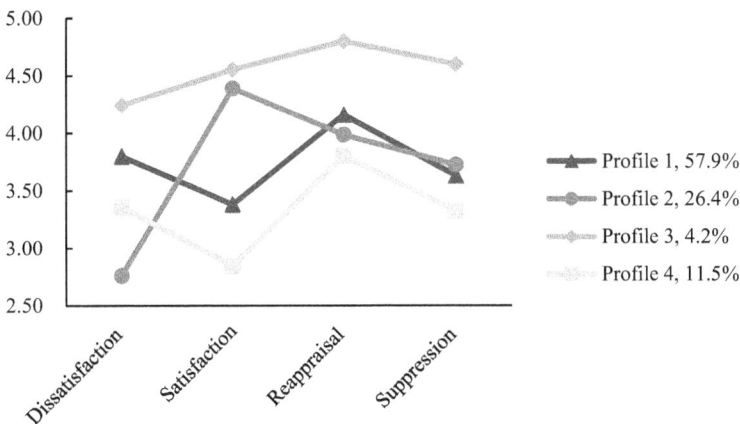

Figure 5.2 Estimated marginal means for dissatisfaction, satisfaction, reappraisal, and suppression by latent profile.

the expressive suppression strategy (F [3, 346] = 3.489, p = .016), which indicated a violation of homogeneity of variances. Therefore, Bonferroni tests were used to compare group differences of satisfaction and dissatisfaction while Games-Howell tests were used to compare differences of the two emotion regulation strategies. The one-way MANOVA revealed a significant multivariate effect of latent profile (F [12, 1035] =14.510, *Pillai's Trace* = 0.432, p < .001). Tests of univariate between-subjects effects indicated significant differences for each dependent variable: Dissatisfaction (F [3, 346] = 24.124, p < .001, η_p^2 = 0.173), satisfaction (F [3, 346] = 39.599, p < .001, η_p^2 = 0.256), reappraisal (F [3, 346] = 5.607, p = .001, η_p^2 = 0.046), and suppression (F [3, 346] = 6.340, p < .001, η_p^2 = 0.052). Post hoc Bonferroni tests indicated that students in latent profile 3 reported significantly higher online dissatisfaction than students in latent profile 4 and profile 2. Both students in latent profiles 1 and 4 had significantly higher online dissatisfaction than latent profile 2. In contrast, latent profile 1 did not significantly differ from latent profile 3 or profile 4 in online dissatisfaction. For online satisfaction, all latent profiles significantly differed from each other except that no significant difference existed between profiles 2 and 3. Regarding the two emotion regulation strategies, Games-Howell tests showed that students in latent profile 3 reported a higher frequency of using the cognitive reappraisal strategy than students in latent profile 4 and profile 2. Simultaneously, students in latent profile 3 reported higher frequency of using the expressive suppression strategy than latent profile 4 and profile 1 in online learning environments.

Discussion

Online learning environments during the COVID-19 pandemic created a whole new experience for teachers' classroom instruction and student learning. The process of adapting to such environments (e.g., adopting new pedagogical strategies or familiarizing themselves with technology tools) has great potential for shaping students' emotional experiences. Using a person-centered approach, this study first analyzed university students' emotional profiles formed by two types of positive emotions (i.e., enjoyment and pride) and two types of negative emotions (i.e., anger and anxiety). Then MANOVA was employed to examine how identified latent emotional profiles related to students' satisfaction and emotion regulation strategies in online learning environments. As I described earlier, these findings make an important contribution to understanding students' achievement emotions in online environments and providing evidence for their theorized relations with other variables, which extends the application of control-value theory in different cultural and learning contexts.

Interpretation of Emotional Profiles

Consistent with previous research conducted in traditional classroom settings, university students experienced diverse emotions in online environments.

The LPA analysis showed that four distinct latent profiles emerged. Each latent profile was characterized by university students' perceived levels of positive emotions (i.e., enjoyment and pride) and negative emotions (i.e., anxiety and anger). Specifically, the four latent profiles were identified as follows: Moderate positive emotions and moderate negative emotions (57.9%); moderately high positive emotions and moderately low negative emotions (26.4%); high positive emotions and high negative emotions (4.2%); low positive emotions and low negative emotions (11.5%). These findings shed light on the different combination of achievement emotions by applying a person-centered approach. Students with moderate positive emotions and moderate negative emotions made up the largest profile while students with high positive emotions and high negative emotions made up the smallest profile.

The distribution of student populations in the four latent profiles reflects the Eastern cultural script that students tend to seek a middle way in Eastern contexts by reaching a balance between positive emotions and negative emotions (Miyamoto & Ma, 2011). This finding contributes to the growing literature by demonstrating the role of cultural scripts in shaping students' emotional experiences. Compared with Western students who are inclined to experience positive emotions frequently and negative emotions infrequently (Eid & Diener, 2001; Kitayama et al., 2000), Eastern students are more likely to experience both positive emotions and negative emotions moderately frequently (Miyamoto & Ryff, 2011). In fact, the dialectical balance of experiencing both positive emotions and negative emotions is associated with healthier profiles for Eastern students (Miyamoto & Ryff, 2011). The second latent profile represented students who could successfully adapt to online learning environments since they perceived moderately high positive emotions and moderately low negative emotions. Based on theoretical assumptions in control-value theory, students in the second profile may achieve more positive learning outcomes than the other three profiles. The third profile represented students who perceived high positive emotions and negative emotions. As such emotional experience or expression is not consistent with the Eastern cultural script, it constitutes the smallest profile as expected (4.2%). Students in this profile typically demonstrate how online learning environments serve as a double-edged sword for student learning (Lee & Choi, 2011). As their experienced emotions are susceptible to external environmental factors, students in the third profile may be the group who are most influenced by the transformation from regular face-to-face classroom instruction to online instruction. The first and third emotional profiles illustrate that emotions with different valances are not completely contradictory among Eastern students. Instead, positive emotions and negative emotions coexist in Eastern learning scenarios. This pattern provides empirical evidence for the previous argument that emotional complexity (i.e., the co-occurrence of positive and negative affect) is more prevalent in Eastern cultural contexts than in Western cultural contexts (Spencer-Rodgers, Peng, & Wang, 2010). The fourth profile represented students who had the lowest positive emotions and lowest negative emotions among four groups. The possible explanation is that these students are relatively calm in online learning environments, either due to the lack of personal relevance of online activities or the consequence of constraining positive emotions and negative emotions. The salient characteristics of four profiles reflect the variability of university students' emotional experiences in online learning environments.

Relations between Students' Emotional Profiles and Outcome Variables

Besides unique characteristics of emotional profiles identified in the university science-major student populations, this study also examined group differences in students' perceived satisfaction with online learning and emotional regulation strategies as a function of profile membership. Results showed that four latent profiles were differentially associated with online satisfaction and emotional regulation strategies. Regarding the variable of online satisfaction, students who reported high satisfaction and low dissatisfaction were more likely to have an emotional profile with moderately high positive emotions and moderately low negative emotions (profile 2). This pattern was consistent with previous research adopting either a variable-centered approach or a person-centered approach (Heckel & Ringeisen, 2019; Parker et al., 2021) and provides additional evidence that achievement emotions are important indicators of student learning experiences in online environments. By contrast, students who perceived more dissatisfaction and less satisfaction in online instruction were more likely to have profiles with low or moderate levels of either positive emotions or negative emotions (profile 1 and profile 4). Consequently, a large proportion of the sample (69.4%) were dissatisfied with the new mode of online instruction. This pattern corroborates the previous argument that emergency remote teaching brings significant challenges for both teachers and students. More attention should be paid to improving teachers' effectiveness in developing and implementing activities in online environments. Compared with the other three latent profiles, students in the third profile who reported high positive emotions and high negative emotions exhibited more complicated attitudes toward online instruction. In other words, average levels of reported online satisfaction and online dissatisfaction were highest among four profiles, which indicated the contradictory or dialectical way of thinking in Eastern contexts. This pattern demonstrates that cultural differences in emotional complexity are mediated by students' beliefs of dialecticism (Spencer-Rodgers, Peng, & Wang, 2010).

Regarding the variable of emotional regulation strategies, Chinese university students employed both the cognitive reappraisal strategy and expressive suppression strategy to regulate their achievement emotions in online learning environments, which was consistent with findings reported in regular classroom settings (Hu et al., 2014). Despite that Eastern cultural values encourage students to control and restrain expressions of intense emotions to retain the group harmony (Soto, Levenson, & Ebling, 2005), university students in this sample were likely to adopt more adaptive emotional regulation strategies (i.e., cognitive reappraisal) and less adaptive emotional regulation strategies (i.e., expressive suppression). Students who reported higher frequencies of using emotional regulation strategies in online learning environments were more likely to have a profile with high positive emotions and high negative emotions (profile 3). It is intriguing to observe that students in profile 3 (high positive emotions and high negative emotions) reported to use the cognitive appraisal strategy more frequently than students in profile 2 (moderately high positive emotions and moderately low negative emotions) and profile 4 (low positive emotions and low negative emotions). This implies that the cognitive appraisal strategy may be used to process negative emotions. Simultaneously, students in profile 3 (high positive emotions and high negative emotions) reported to use the expressive suppressive strategy more frequently than students in profile 1 (moderate positive emotions and moderate negative emotions) and profile 4 (low positive emotions and low negative emotions), which implies that the expressive suppressive strategy may be used to process both positive emotions and negative emotions. However, results regarding relations between the expression of positive or negative emotions and the use of different emotion regulation strategies should be interpreted with caution since the sample size of profile 3 is relatively small.

Conclusions and Implications

The outbreak of the COVID-19 pandemic has changed the way of school instruction and led to positive or negative effects on students' learning experiences. Based on the control-value theory, this study first examined how science-major university students' enjoyment, pride, anxiety, and anger comprised four different emotional profiles in online learning environments, namely moderate positive emotions and moderate negative emotions (profile 1), moderately high positive emotions and moderately low negative emotions (profile 2), high positive emotions and high negative emotions (profile 3), and low positive emotions and low negative emotions (profile 4). Then their relations with online satisfaction and emotional regulation strategies were further investigated. Compared with the other three groups, our findings showed that students who reported moderately high positive emotions and moderately low negative emotions were more satisfied with online instruction. Simultaneously, students who reported high positive emotions and high negative emotions more frequently used expressive suppression and cognitive reappraisal regulation strategies.

Considering the significant effect of achievement emotions on students' learning behaviors and outcomes, findings of this study point to the necessity of paying attention to students' emotional experiences in online learning environments. It also provides implications for developing pedagogical interventions that create emotion-friendly online learning environments. As online learning environments are less structured than traditional classroom settings due to reduced face-to-face interactions, teachers may intentionally create opportunities for supporting and monitoring students in the following pathways. First, teachers may spend time identifying and solving students' technical problems at the beginning of using new technology tools. Second, teachers may develop small online assessment tasks to enhance communication with students and monitor student progress during class time. Third, teachers may encourage students to reflect and improve their online learning strategies and provide instant and personalized feedback after online classes. In addition, students' positive emotions might be fostered if online activities and tasks are developed in different formats, which allows students to retain engaged, collaborate with peers, and achieve conceptual understanding through various learning pathways. To achieve these goals, teacher education and professional development programs may help pre-service or in-service teachers to structure online activities in appropriate ways targeting to enhance students' positive emotions or reduce negative emotions, and finally improve the effectiveness of online instruction.

Limitations and Future Directions

As with all research, several limitations should be considered when interpreting results of this study. First, there were only 15 students in profile 3, which represented 4.2% of the sample population. It is recommended that ideal models should contain few profiles that comprise less than 5% of the total sample (DiStefano & Kamphaus, 2006). As students in profile 3 showed the pattern of high positive emotions and high negative emotions, its small proportion is theoretically explainable. Yet, I suggest future research may collect data with larger sample sizes to construct more accurate models of students' emotional profiles. Second, only self-report survey data were used in the data analysis. The potential concern is that Chinese students may provide more moderate responses on the rating scale (Chen & Stevenson, 1995). Such effect may be aggravated or alleviated since some students might underestimate or overestimate their perceptions of positive or negative emotions due to "a retrospective recall bias". Therefore, I suggest supplementing survey data with open-ended responses or interview data to investigate details or mechanisms of how achievement emotions and emotional regulation strategies are related in different profiles. Third, participants in this study were recruited from science-major juniors or sophomores who enrolled in a teacher education course. Results of this study only reflect online learning experiences of this particular sample and cannot be generalized to science-major students in general. Consequently, I recommend that researchers can collect survey data from more diverse student populations to fully understand emotions and regulation strategies in online learning environments.

References

Aldao, A. (2013). The future of emotion regulation research: Capturing context. *Perspectives on Psychological Science, 8*, 155–172. http://dx.doi.org/10.1177/1745691612459518

Arbaugh, J. B. (2000). Virtual classroom characteristics and student satisfaction with internet-based MBA courses. *Journal of Management Education, 24*(1), 32–54.

Artino, A. R., & Jones, K. D. (2012). Exploring the complex relations between achievement emotions and self-regulated learning behaviors in online learning. *Internet and Higher Education, 15*(3), 170–175. https://doi.org/10.1016/j.iheduc.2012.01.006

Barrett, L. F., & Russell, J. A. (1998). Independence and bipolarity in the structure of current affect. *Journal of Personality and Social Psychology, 74*(4), 967–984. https://doi.org/10.1037/0022-3514.74.4.967

Buhr, E. E., Daniels, L. M., & Goegan, L. D. (2019). Cognitive appraisals mediate relationships between two basic psychological needs and emotions in a massive open online course. *Computers in Human Behavior, 96*, 85–94.

Butler, E. A., & Gross, J. J. (2009). Emotion and emotion regulation: Integrating individual and social levels of analysis. *Emotion Review, 1*, 86–87.

Butler, E. A., Lee, T. L., & Gross, J. J. (2007). Emotion regulation and culture: Are social consequences of emotion suppression culture specific? *Emotion, 7*, 30–48.

Butler, E. A., Lee, T. L., & Gross, J. J. (2009). Does expressing your emotions raise or lower your blood pressure? The answer depends on cultural context. *Journal of Cross-Cultural Psychology, 40*, 510–517.

Chen, C., & Stevenson, H. W. (1995). Motivation and mathematics achievement: A comparative study of Asian-American, Caucasian-American, and East Asian high school students. *Child Development, 66*(4), 1215–1234.

Deng, W., Lei, W., Guo, X., Li, X., Ge, W., & Hu, W. (2022). Effects of regulatory focus on online learning engagement of high school students: The mediating role of self-efficacy and academic emotions. *Journal of Computer Assisted Learning, 38*(3), 707–718.

Ding, Y., & Zhao, T. (2020). Emotions, engagement, and self-perceived achievement in a small private online course. *Journal of Computer Assisted Learning, 36*(4), 449–457.

DiStefano, C., & Kamphaus, R. W. (2006). Investigating subtypes of child development: A comparison of cluster analysis and latent class cluster analysis in typology creation. *Educational and Psychological Measurement, 66*(5), 778–794.

Dryman, M. T., & Heimberg, R. G. (2018). Emotion regulation in social anxiety and depression: a systematic review of expressive suppression and cognitive reappraisal. *Clinical Psychology Review, 65*, 17–42.

Eid, M., & Diener, E. (2001). Norms for experiencing emotions in different cultures: Inter- and intranational differences. *Journal of Personality and Social Psychology, 81*, 869–885.

Gaudreau, P., Miranda, D., & Gareau, A. (2014). Canadian university students in wireless classrooms: What do they do on their laptops and does it really matter? *Computers and Education, 70*, 245–255.

Gross, J. J. (2013). Emotion regulation: Taking stock and moving forward. *Emotion, 13*, 359–365. http://dx.doi.org/10.1037/a0032135

Gross, J. J. (2014). Emotion regulation: Conceptual and empirical foundations. In J. J. Gross (Ed.), *Handbook of emotion regulation* (2nd ed., pp. 3–22). New York: Guilford Press.

Gross, J. J. (2015). The extended process model of emotion regulation: Elaborations, applications, and future directions. *Psychological Inquiry, 26*(1), 130–137. https://doi.org/10.1080/1047840X.2015.989751

Gross, J. J., & Jazaieri, H. (2014). Emotion, emotion regulation, and psychopathology: An affective science perspective. *Clinical Psychological Science, 2*(4), 387–401. https://doi.org/10.1177/2167702614536164

Gross, J. J., & John, O. P. (2003). Individual differences in two emotion regulation processes: Implications for affect, relationships, and well-being. *Journal of Personality and Social Psychology*, *85*(2), 348–362.

Gross, J. J., Sheppes, G., & Urry, H. L. (2011). Emotion generation and emotion regulation: A distinction we should make (carefully). *Cognition & Emotion*, *25*, 765–781.

Hagenaars, J. A., & McCutcheon, A. L. (2002). *Applied latent class analysis* (J. A. Hagenaars & A. L. McCutcheon (Eds.). Cambridge, UK: Cambridge University Press.

Heckel, C., & Ringeisen, T. (2019). Pride and anxiety in online learning environments: Achievement emotions as mediators between learners' characteristics and learning outcomes. *Journal of Computer Assisted Learning*, *35*(5), 667–677. https://doi.org/10.1111/jcal.12367

Hickendorff, M., Edelsbrunner, P. A., McMullen, J., Schneider, M., & Trezise, K. (2018). Informative tools for characterizing individual differences in learning: Latent class, latent profile, and latent transition analysis. *Learning and Individual Differences*, *66*, 4–15.

Horesh, D., & Brown, A. D. (2020). Traumatic stress in the age of COVID-19: A call to close critical gaps and adapt to new realities. *Psychological Trauma: Theory, Research, Practice, and Policy*, *12*(4), 331–335.

Hu, T., Zhang, D., Wang, J., Mistry, R., Ran, G., & Wang, X. (2014). Relation between emotion regulation and mental health: A meta-analysis review. *Psychological Reports*, *114*(2), 341–362. https://doi.org/10.2466/03.20.PR0.114k22w4

John, O. P., & Gross, J. J. (2004). Healthy and unhealthy emotion regulation: Personality processes, individual differences, and life span development. *Journal of Personality*, *72*(6), 1301–1334.

Kecojevic, A., Basch, C. H., Sullivan, M., & Davi, N. K. (2020). The impact of the COVID-19 epidemic on mental health of undergraduate students in New Jersey, cross-sectional study. *PloS one*, *15*(9), Article e0239696.

Kitayama, S., Markus, H. R., & Kurokawa, M. (2000). Culture, emotion, and well-being: Good feelings in Japan and the United States. *Cognition and Emotion*, *14*(1), 93–124. https://doi.org/10.1080/026999300379003

Kline, R. B. (1998). *Principles and practice of structural equation modeling*. New York, NY: Guilford Press.

Lazarus, R. S., & Alfert, E. (1964). Short-circuiting of threat by experimentally altering cognitive appraisal. *Journal of Abnormal and Social Psychology*, *69*, 195–205.

Lee, J. (2020). Mental health effects of school closures during COVID-19. *The Lancet Child & Adolescent Health*, *4*(6), 421.

Lee, M., Pekrun, R., Taxer, J. L., Schutz, P. A., Vogl, E., & Xie, X. (2016). Teachers' emotions and emotion management: Integrating emotion regulation theory with emotional labor research. *Social Psychology of Education*, *19*(4), 843–863. https://doi.org/10.1007/s11218-016-9359-5

Lee, Y., & Choi, J. (2011). A review of online course dropout research: Implications for practice and future research. *Educational Technology Research and Development*, *59*(5), 593–618.

Liu, Y., & Sang, B. (2020). Academic emotional expression in middle school students and its relation with academic emotion. *Journal of Psychological Science*, *43*(3), 600–607. (In Chinese)

Lo, Y., Mendell, N. R., & Rubin, D. B. (2001). Testing the number of components in a normal mixture. *Biometrika*, *88*(3), 767–777.

Miyamoto, Y., & Ma, X. (2011). Dampening or savoring positive emotions: A dialectical cultural script guides emotion regulation. *Emotion, 11*(6), 1346–1357. https://doi.org/10.1037/a0025135

Miyamoto, Y. & Ryff, C. (2011). Cultural differences in the dialectical and non-dialectical emotional styles and their implications for health. *Cognition and Emotion, 25*, 22–30.

Moore, M., & Kearsley, G. (2012). *Distance education: A systems view of online learning* (3rd ed.). Belmont, CA: Wadsworth.

Nylund-Gibson, K., Grimm, R., Quirk, M., & Furlong, M. (2014). A latent transition mixture model using the three-step specification. *Structural Equation Modeling: A Multidisciplinary Journal, 21*(3), 439–454.

Oyserman, D., Coon, H. M., & Kemmelmeier, M. (2002). Rethinking individualism and collectivism: Evaluation of theoretical assumptions and meta-analyses. *Psychological Bulletin, 128*, 3–72.

Parker, P. C., Perry, R. P., Hamm, J. M., Chipperfield, J. G., Pekrun, R., Dryden, R. P., Daniels, L. M., & Tze, V. M. C. (2021). A motivation perspective on achievement appraisals, emotions, and performance in an online learning environment. *International Journal of Educational Research, 108*(November), 101772. https://doi.org/10.1016/j.ijer.2021.101772

Pekrun, R. (2000). A social cognitive, control–value theory of achievement emotions. In J. Heckhausen (Ed.), *Motivational psychology of human development* (pp. 143–163). Oxford, England: Elsevier.

Pekrun, R. (2006). The control-value theory of achievement emotions: Assumptions, corollaries, and implications for educational research and practice. *Educational Psychology Review, 18*(4), 315–341.

Pekrun, R., Goetz, T., & Frenzel, A. C. (2005). *Academic Emotions Questionnaire – Mathematics (AEQ-M): User's manual.* Munich: University of Munich, Department of Psychology.

Pekrun, R., Goetz, T., Titz, W., & Perry, R. P. (2002). Academic emotions in students' self-regulated learning and achievement: A program of qualitative and quantitative research. *Educational Psychologist, 37*(2), 91–105.

Racanello, D., Balbontín-Alvarado, R., Bezerra, D., Burro, R., Cheraghi, M., Dobrowolska, B., … Aristovnik, A. (2022). Higher education students' achievement emotions and their antecedents in e-learning amid COVID-19 pandemic: A multi-country survey. *Learning and Instruction, 80*, 101629.

Richardson, J. T. (2011). Eta squared and partial eta squared as measures of effect size in educational research. *Educational Research Review, 6*(2), 135–147.

Schwarz, G. (1978). Estimating the dimension of a model. *The Annals of Statistics, 6*(2), 461–464.

Sorić, I., Penezić, Z., & Burić, I. (2013). Big Five personality traits, cognitive appraisals and emotion regulation strategies as predictors of achievement emotions. *Psihologijske Teme, 22*(2), 325–349.

Soto, J. A., Levenson, R. W., & Ebling, R. (2005). Cultures of moderation and expression: emotional experience, behavior, and physiology in Chinese Americans and Mexican Americans. *Emotion, 5*(2), 154.

Soto, J. A., Perez, C. R., Kim, Y. H., Lee, E. A., & Minnick, M. R. (2011). Is expressive suppression always associated with poorer psychological functioning? A cross-cultural comparison between European Americans and Hong Kong Chinese. *Emotion, 11*(6), 1450–1455. https://doi.org/10.1037/a0023340

Spencer-Rodgers, J., Peng, K., & Wang, L. (2010). Dialecticism and the co-occurrence of positive and negative emotions across cultures. *Journal of Cross-Cultural Psychology*, *41*(1), 109–115.

Stephan, M., Markus, S., & Gläser-Zikuda, M. (2019). Students' achievement emotions and online learning in teacher education. *Frontiers in Education*, *4*(October), 1–12. https://doi.org/10.3389/feduc.2019.00109

Wormington, S. V., & Linnenbrink-Garcia, L. (2017). A new look at multiple goal pursuit: The promise of a person-centered approach. *Educational Psychology Review*, *29*(3), 407–445.

You, J. W., & Kang, M. (2014). The role of academic emotions in the relationship between perceived academic control and self-regulated learning in online learning. *Computers & Education*, *77*, 125–133.

Part 4

The Application of Situated Expectancy-Value Theory in Chinese Science Classrooms

6 The Development and Validation of the Chemistry Value Questionnaire

Introduction

Subjective values, defined as perceived values or reasons for engaging in tasks, are important predictors of individuals' academic choices, persistence, and performance in STEM (Science, Technology, Engineering, and Mathematics) disciplines (Eccles et al., 1983; Wigfield & Eccles, 2020; Wang, & Degol, 2013). Despite the fact that students often acknowledge the importance of science subjects in promoting social and economic development, they often think school science education is content-dominated, difficult, and uninteresting due to the lack of relevance to daily lives (Osborne & Collins, 2001). This phenomenon is more pronounced for the chemistry subject because it involves many unobservable microscopic structures and mechanisms. In other words, there exist gaps between educators' views and students' views of values of learning science. For example, the notion of the value of learning science is often highlighted in Next Generation Science Standards (NGSS) (National Research Council, 2012) and Chinese Science Curriculum Standards (CSCS) (Ministry of Education, 2020). The NGSS document encourages teachers to develop students' eight science practices (e.g., developing and using models and engaging in argument from evidence) while the CSCS document describes that the curriculum goal of compulsory science education is to develop students' core literacies including content knowledge, key skills, and correct values (e.g., ethical and moral values). These competences covered in the aforementioned literacies enable students to make the right decisions in the face of social scientific issues (e.g., scientific research and technology application). Specifically, the value of learning chemistry is demonstrated in the following curriculum standards: Students can illustrate the value of applying chemical reactions in nature, industry, and daily lives and acknowledge its unique value of benefiting human beings. In addition, students may form the value of shouldering the responsibility of protecting the environment and pursuing the harmony between humans and nature. Considering the central role of value in science curriculum standards and its cultural or domain specificity, it is worthwhile to conceptualize the construct of value and investigate how students perceive values of learning chemistry in Chinese school contexts.

DOI: 10.4324/9781032649665-10

Literature Review

Before developing instrument items, I reviewed previous literature about how to define and measure the value construct in science education and education psychology fields. In this section, I first described the conceptualization of the value construct, then listed existing surveys measuring this construct and finally systematically described how the instrument was developed.

The value of learning science is often represented by its degree of personal relevance. Eilks and Hofstein (2015) suggested that chemistry teachers should make chemistry teaching and learning more relevant during the development and implementation of the chemistry curriculum. In this book, Stuckey, Hofstein, Mamlok-Naaman, and Eilks (2013) provided a three-dimensional model for elaborating the definition of relevance in science education. As shown in Table 6.1, the relevance of learning science is represented at individual dimension, vocational dimension, and societal dimension. These different types of values, which are embodied in different time periods (i.e., at present or in the future), cover both intrinsic and extrinsic components. For example, the contribution of learning science in getting good and well-paid jobs is related to future vocational goals. This model provides the theoretical framework for understanding the conceptualization of values in science education. However, as different types of values are intertwined in nature, boundaries between three different levels are somewhat vague (Gong, Bergey, Jin, Mao, & Cheng, 2023). What is more, the science curriculum is embedded with rich inquiry activities. Allchin (1998) argued that epistemic value is represented in scientific methods and research which includes controlled observations,

Table 6.1 Stuckey et al.'s model of three dimensions of the relevance of science education (2013)

	Intrinsic Value		Extrinsic Value	
	Present	Future	Present	Future
Individual dimension	Satisfying curiosity and interest	Skills needed for future life	Getting high grades	Acting responsibly and solidaric in future
Societal dimension	Finding one's own place in society	Promoting one's own interests in societal discourse	Learning how to behave in society	Behaving as a responsible citizen
Vocational dimension	Orientation about potential careers	Getting good and well-paid jobs	Passing exams qualifying for coming education	Contributing to society's economic growth

confirmation of predications, repeatability, and so on. Corrigan, Cooper, and Keast (2015) corroborated the preceding viewpoint and proposed that epistemic values underlying chemistry discipline include curiosity, creativity, open-mindedness, rational thinking, empiricism, skepticism, and accuracy. Therefore, besides intrinsic value and extrinsic value at three dimensions, epistemic value should also be taken into consideration when developing survey items.

Besides Stuckey et al.'s model, existing surveys measuring students' perceived values of learning one specific subject usually adopt situated expectancy-value theory (SEVT) as the theoretical framework. Eccles (2005) outlined task values in terms of three positive components: Intrinsic value (i.e., interest), attainment value (i.e., importance), and utility value. Intrinsic value is defined as inherent interests or enjoyment toward the subject. Attainment value is defined as the personal importance of doing well on the task and is closely related to the individual's self-schema (Wigfield, Tonks, & Klauda, 2009). Utility value is defined as the perceived usefulness of tasks for achieving short-term or long-term goals. Recent research has included a negative component – cost –because students may perceive math as less valuable if the costs outweigh benefits. It is defined as the subjective evaluation of loss due to doing the task. For example, Luttrell et al. (2010) developed the Mathematics Value Inventory (MVI) to measure general education students' perceived value of mathematical literacy. Items covered students' beliefs in four areas: Interest, general utility, need for high achievement, and personal cost. Interest is similar to intrinsic value. General utility is defined as the value of understanding math in accomplishing personal goals. The need for high achievement is the perceived importance of performing well in math and belongs to attainment value. Gaspard et al. (2020) have made finer distinctions for utility value, including utility for daily life, job, or school and social utility, or the utility for retaining relations with peers. These different types of values in SEVT overlap with intrinsic value and extrinsic value in the individual dimension and vocational dimension described in Stuckey et al.'s model.

As values are not isolated from cultural, social, and disciplinary contexts, our research team has explored how the notion of value is represented in Chinese chemistry curriculum standards and school textbooks. To address this issue, we selected the two typical units in high school chemistry curriculum and developed two surveys measuring how students perceived the value of learning the mentioned two units (Luo, 2021; Shi, 2021). One unit is titled "Iron and metallic materials" while the other unit is titled "Organic compounds". Before developing survey items, qualitative content analysis was used to examine how the value construct was represented in curriculum standards and school textbooks related to these units. The related information was extracted for developing survey items. Then exploratory factor analyses were adopted to examine internal structures of the two surveys. Regarding the first unit, as shown in Table 6.2 (Shi, 2021), students' perceived values of learning the unit of iron and metallic materials were composed of four factors: The first factor – social

Table 6.2 Survey items measuring students' perceived values of learning the unit of iron and metallic materials (Shi, 2021)

Factors	Survey Items
Social Value and Daily-Life Utility Value	• Learning this unit can help me understand the influence of minor elements on human health and pay attention to food selection. • Learning this unit helps me pay attention to the utilization and protection of rare earth resources. • I can use the knowledge learned to get involved in the discussion of social scientific issues (e.g., utilize and protect natural resources). • I can identify whether products in daily lives are made of iron, aluminum, or alloy. • Learning this unit helps me understand how iron and other metals contribute to social development.
Epistemic Value	• Learning this unit can help me understand how metal elements transform between different valence states. • Learning this unit helps me understand properties of iron and related chemical compounds. • Learning the application of metal and alloy can help me understand the close connection between chemistry and daily lives. • Learning this unit helps me infer properties of substances based on the substance's category and the element's valence.
Vocational Utility Value	• Learning this unit increases my interest in choosing chemistry-related jobs (e.g., test engineer). • Learning this unit promotes me to select chemistry as the subject tested in the national college examination.
Problem-Solving Utility Value	• Learning this unit helps me analyze phenomena or mechanisms in labs, industry, and daily lives based on properties of different substances. • Learning this unit helps me solve chemistry-related problems in daily lives (e.g., measuring the iron element in food samples).

value and daily life utility value – described how metallic elements were related to daily lives, social development, resource management, and environmental protection; the second factor – epistemic value – described how principles of the variation of metal valence helped students make inferences; the third factor – vocational utility value – described its value for future academic or occupational choices; and the fourth factor – problem solving utility value – was related to the value of solving chemistry-related problems. It should be noted that problem-solving scenarios in the fourth factor were more complicated than those described in the first factor. Regarding the organic-compound unit, as shown in Table 6.3 (Luo, 2021), three factors emerged: The first

Table 6.3 Survey items measuring students' perceived values of learning the unit of organic compounds (Luo, 2021)

Factors	Survey Items
Epistemic Value	• Learning this unit can help me understand properties of organic compounds' functional groups.
	• I can use the phenomena observed in experiments as the evidence to infer properties of organic compounds.
	• Learning this unit can help me understand properties of common organic compounds.
	• Ball-and-stick models can deepen my understanding of the structure of organic compounds.
	• Using methane as an example to summarize alkanes' properties can help me make generalizations.
	• Organic-chemistry experiments can improve my scientific inquiry skills.
	• Learning the knowledge of basic nutrients from food can help me understand the connection between chemistry and biology.
Social and Intrinsic Value	• Learning this unit helps me understand that knowledge of organic chemistry is widely applied to chemical production.
	• Learning knowledge of organic polymer materials helps me understand the important role of organic chemistry in advanced industries (e.g., the spaceflight and shipping industry).
	• Learning this unit helps me understand that knowledge of organic compounds is widely applied to daily lives.
	• Learning this unit helps me understand that many objects in daily lives are composed of chemical substances (e.g., the food package is made of polyethylene).
	• Our scientists' contributions in the organic chemistry area make me feel proud.
	• Knowing the property of organic compounds can help us avoid chemical accidents in industry.
	• The process of identifying benzene's structure helps me feel the scientific spirit of pursuing truth.
Utility Value	• Inquiry chemistry experiments in this unit increase my interest in learning chemistry.
	• Learning this unit helps me grasp and apply the method of experiment design.
	• I can transfer knowledge of organic compounds into other areas (e.g., compare the acidity of organic acids or describe the structure of glycine and glutamic acid).
	• Learning knowledge of organic compounds can help me be involved in discussing social scientific issues (waste incineration).
	• The summary of general steps of understanding organic compounds in the textbook can help me analyze new organic compounds independently.
	• Learning knowledge of organic compounds can help me explain phenomena and solve problems in daily life.

factor – epistemic value – described how content knowledge in this unit helped students understand structures and properties of organic compounds, and make inferences based on organic compounds' functional groups; the second factor – social and intrinsic value – described the important role of organic chemistry in industry and how role models incited students' feeling of pride and scientific spirits; the third factor – utility value – mainly described how students could apply or transfer knowledge to new or complex problems. These studies have demonstrated that each unit has its own unique characteristics of cultivating students' ways of thinking and perceived values of learning chemistry, which provides rich information for the conceptualization of values of learning chemistry.

The Current Study

The conceptualization of values depends on the subject domain under consideration. Different ways of conceptualizing the value construct in education psychology (i.e., SEVT) and science education fields (i.e., Stuckey et al.'s model) demonstrate that additional research is needed to investigate students' perceived values of learning chemistry. In this section, students' open responses were first collected to examine potential categories of perceived values of learning chemistry in Chinese high school contexts, which provided rich resources for constructing initial items, along with existing validated surveys. Then the confirmatory factor analysis was adopted to examine the survey's internal structure.

Method for Content Validation

Before developing survey items, I collected high students' responses to an open-format question: "What is the value of learning chemistry?" The purpose of this step was to gather evidence for examining whether and how students' value perceptions matched with existing categories in the literature. The data was collected in the fall semester during the 2019–2020 academic year. The sample was 941 10th-grade and 11th-grade students from two private high schools located in a northern province (M = 15.69 years old, SD = 0.90). Among them, 46.55% were female, 50.37% were tenth grade, and 3.0% were from minority groups. The survey collection was administered by schoolteachers and it took students about 15 minutes to complete the paper survey. Then the qualitative content analysis (Braun & Clarke, 2006) was adopted to identify various categories of students' perceived value of learning chemistry.

Based on the conceptualization of the value construct in SEVT and Stuckey et al.'s model, six types of perceived values of learning chemistry was identified in students' open responses (see Table 6.4): Utility value, intrinsic value,

Table 6.4 Six types of values identified in students' open responses

Types	Codes	Examples of Students' Open Responses
Utility Value	Academic	• I learn it for exams. • It is very important in the national college entrance exam. • It can help me learn biology.
	Job	• It can help me have more options when selecting jobs. • I may choose chemistry-related jobs in the future. • It is related to the university major that I want to learn. • It helps me learn medicine and do scientific research in the future.
	Practical	• I can use chemistry knowledge to explain magic tricks. • It contributes to understanding some phenomena in daily lives. • At least it lets me know what is dangerous to eat. • It helps me understand ingredients of medicines and skin care products. • It can help me avoid buying pseudo-salt and pseudo-oil, use disinfectants correctly, and avoid fraud or false advertising.
	Relation	• I can do experiments to do a performance for my classmates. • I can brag to classmates about my chemistry knowledge. • I want to learn chemistry well and not disappoint others.
	Research	• It enables me to use experiments to explore the truth. • It enables me to pursue the truth by analyzing problems from different perspectives. • It can help us to produce high-quality new materials.
Intrinsic Value		• Chemistry is interesting. I feel happy when learning chemistry. • I just want to learn it! • Chemistry experiments are interesting. Learning chemistry is the process of spending effort on one's own interests.
Attainment Value		• I like digging deep into things. Learning chemistry meets my needs. • It shows the value of life. • It is very significant for improving oneself. • Learning chemistry makes me feel a sense of accomplishment. • Learning it well demonstrates my intelligence.
Social Value		• Learning chemistry enables me to contribute to the country and society. • It can help human beings to utilize earth resources more efficiently and maximize social values. • Chemistry can make our society and life more convenient and improve the environment.
Epistemic Value		• It can enrich my knowledge and broaden my horizons. • It helps me grasp the skill of rational thinking and logical thinking. • It can help us make connections between the macroscopic level and the microscopic level.
Aesthetic Value		• It shows nature's magic and beauty. • Chemistry is the most beautiful subject in the world.

attainment value, social value, epistemic value, and aesthetic value. Among them, utility value, intrinsic value, and attainment value were consistent with those subcategories defined in SEVT while social value was related to the social dimension in Stuckey et al.'s model. Specifically, utility value covered a wide range of values including **academic utility** (getting good grades and learning other subjects), **practical utility** (explaining phenomena and making right decisions in daily lives), **occupational utility** (having more job opportunities and selecting university majors), **relational utility** (maintaining relations with classmates, parents and teachers), and **research utility** (discovering new things and pursuing the truth). Intrinsic value referred to personal initiatives, interests, and positive feelings (e.g., happiness) toward the subject of chemistry. The attainment value was represented in formats of meeting students' needs of self-fulfillment and demonstrating their intelligence. The social value was related to how the student self utilized chemistry knowledge to contribute to human beings, countries, societies, and natural environments. The preceding social value reflects the collectivist cultural norms that individuals may fulfill their social obligations to maintain group harmony (Dreamson, 2018). The epistemic value included the general benefit (e.g., enriching knowledge and broadening horizon) and domain-specific benefits (e.g., grasping rational thinking and connecting between macroscopic, microscopic, and symbolic levels). In addition, aesthetic value defined as the "deep appreciation for the beauty and power of science ideas that transform one's experiences and perceptions of the world" (Girod, Rau, & Schepige, 2003, p.577),was identified because chemistry textbooks include various photographs of chemistry-related natural phenomena, chemical reactions, and models of molecule structures.

The Instrument Validity

Based on students' responses and existing survey items in previous literature, I developed a self-report instrument in a 5-point Likert-type response format, with the following response options: 1 (*strongly disagree*), 2 (*disagree*), 3 (*neutral*), 4 (*agree*), and 5 (*strongly agree*). The instrument included 32 items and measured students' perceived values of learning chemistry in six broad categories: Intrinsic value, attainment value, utility value, epistemic value, aesthetic value, and social value. Among them, items measuring intrinsic value, attainment value, and utility value were adapted from the Mathematics Value Inventory (MVI; Luttrell et al., 2010) and/or TIMSS2019 while items measuring epistemic value, aesthetic value, and social value were researcher-developed based on previous research (Luo, 2021; Nieswandt, 2007; Shi, 2021) and Chinese chemistry curriculum standards. Full details of survey items are listed in Table 6.5.

Table 6.5 Survey items for measuring six types of values

Categories	Items
Intrinsic Value (N = 4)	Q1: I enjoy learning chemistry.
	Q2: I find many topics in chemistry to be interesting.
	Q3: Chemistry fascinates me.
	Q4: I am interested in solving chemistry problems.
Attainment Value (N = 5)	Q1: Getting high grades in chemistry is important to me.
	Q2: Understanding the subject content of chemistry is important to me.
	Q3: Doing well in chemistry can fulfill parents' expectations.
	Q4: If I do well in chemistry, I can help my classmates with a sense of fulfillment.
	Q5: It is worthwhile to spend effort in doing well in chemistry.
Utility Value (N = 5)	Q1: I need chemistry to learn other school subjects.
	Q2: I think learning chemistry will help me get good grades in exams.
	Q3: Learning chemistry will benefit my choice of university major in the future.
	Q4: Learning chemistry will give me more job opportunities.
	Q5: I think learning chemistry will help me solve problems in my daily life.
Epistemic Value (N = 9)	Q1: Learning chemistry helps me think rigorously.
	Q2: Learning chemistry can increase my safety awareness.
	Q3: Learning chemistry can develop my creativity.
	Q4: Learning chemistry can strengthen my scientific inquiry skills.
	Q5: Learning chemistry helps me make inferences informed by evidence.
	Q6: Learning chemistry helps me understand how to examine the scientific basis behind phenomena.
	Q7: Learning chemistry helps me treat things critically.
	Q8: Learning chemistry enables me to participate in the discussion and practice of related societal issues.
	Q9: Learning chemistry enables me to solve problems by identifying patterns and applying principles.
Aesthetic Value (N = 4)	Q1: When learning chemistry, I can feel the beauty of rationality.
	Q2: When learning chemistry, I can feel the beauty of conservation.
	Q3: When learning chemistry, I can feel the beauty of diverse phenomena.
	Q4: When learning chemistry, I can feel the beauty of symmetry.
Social Value (N = 5)	Q1: Learning chemistry can help us make economic and social progress.
	Q2: Learning chemistry can help humans keep healthy.
	Q3: Learning chemistry can help us protect the environment and achieve sustainable development.
	Q4: Learning chemistry can help improve our quality of life.
	Q5: Learning chemistry can help provide us with new materials and energy sources.

Confirmatory factor analysis is commonly used to examine theoretically hypothesized relations between items and latent factors. In this study, confirmatory factor analysis using maximum likelihood estimation was conducted to examine its internal structure validity in Mplus. Based on theoretical constructs in situated expectancy-value theory and qualitative content analysis of students' open responses, I tested a six-factor model (i.e., intrinsic value, attainment value, utility value, social value, epistemic value, and aesthetic value) with the survey data collected from 464 10th-grade and 11th-grade students (M = 15.66 years old, SD = 0.73; 43.6% female; 3.3% minority) from a private high school in a northern province in China. Model fit was examined by following indexes, including the standardized root mean square residual (SRMR), the root mean square error of approximation (RMSEA), comparative fit index (CFI), and Tucker Lewis index (TLI). Based on Hu and Bentler's (1999) criteria, the six-factor solution produced an adequate model fit (SRMR = .06 < .08; RMSEA= .06 < .08; CFI = .91 > .90; TLI = 0.90) while the one-factor solution did not fit the data well (SRMR = .09; RMSEA= .12; CFI = .66; TLI = .63) (see Table 6.6). What is more, the six-factor solution had lower AIC and BIC values (AIC = 36348.32; BIC = 36807.13) than the one-factor solution (AIC = 39814.61; BIC = 40211.41), which provided evidence for the construct validity. Cronbach's alpha for the whole instrument was 0.969 (N = 32). Cronbach's alphas for intrinsic value, attainment value, utility value, social value, epistemic value, and aesthetic value were 0.951, 0.834, 0.850, 0.939, 0.953, and 0.927. The standardized confirmatory factor loadings are presented in Table 6.7. Loadings ranged between 0.615 and 0.944.

Descriptive statistics and correlations among six factors are listed in Table 6.8. High school students perceived the highest level of social value (*M* = 5.21; *SD* = 0.94), followed by epistemic value (*M* = 4.82; *SD* = 1.02). Means of intrinsic value, utility value, attainment value, and aesthetic value were relatively similar. Bivariate correlation results supported theorized relations: Six different types of values were positively and significantly correlated with each other, ranging in magnitude from 0.39 (between intrinsic value and social value) to 0.79 (between epistemic value and aesthetic value). Among them, intrinsic value had the strongest correlation with students' chemistry scores, while social value had the weakest correlation.

Table 6.6 Goodness-of-fit indices for two models

Model	χ^2	*p*	*CFI*	*TLI*	*SRMR*	*RMSEA*	*AIC*	*BIC*
One-factor	3311.16	<0.001	.66	.63	.09	.12	39814.61	40211.41
Six-factor	1209.63	<0.001	.91	.90	.06	.06	36348.32	36807.13

Table 6.7 Confirmatory factor analysis output for value items

Factors	Items	Standardized Factor Loadings	S.E.
Intrinsic Value (IV)	IV1	0.939	0.010
	IV2	0.911	0.012
	IV3	0.926	0.016
	IV4	0.876	0.017
Utility Value (UV)	UV1	0.694	0.056
	UV2	0.623	0.045
	UV3	0.863	0.045
	UV4	0.848	0.048
	UV5	0.621	0.063
Attainment Value (AtV)	AtV1	0.734	0.041
	AtV2	0.815	0.028
	AtV3	0.615	0.041
	AtV4	0.688	0.034
	AtV5	0.687	0.043
Aesthetic Value (AeV)	AeV1	0.855	0.032
	AeV2	0.821	0.034
	AeV3	0.944	0.013
	AeV4	0.870	0.027
Epistemic Value (EV)	EV1	0.877	0.016
	EV2	0.792	0.031
	EV3	0.823	0.029
	EV4	0.878	0.020
	EV5	0.860	0.023
	EV6	0.870	0.019
	EV7	0.833	0.026
	EV8	0.751	0.028
	EV9	0.810	0.026
Social Value (SV)	SV1	0.786	0.032
	SV2	0.839	0.032
	SV3	0.916	0.020
	SV4	0.904	0.034
	SV5	0.909	0.019

Table 6.8 Descriptive statistics and bivariate correlations among six types of values

	IV	UV	AtV	AeV	EV	Score	M	SD
Intrinsic Value (IV)						.49	4.52	1.34
Utility Value (UV)	.75					.38	4.54	1.12
Attainment Value (AtV)	.66	.71				.33	4.57	1.10
Aesthetic Value (AeV)	.70	.61	.64			.39	4.50	1.24
Epistemic Value (EV)	.65	.66	.63	.79		.36	4.82	1.02
Social Value (SV)	.39	.49	.47	.55	.71	.26	5.21	0.94

Note: all correlations were significant at $p < 0.01$.

Conclusions and Discussion

In this chapter, I systematically described the development and validation of an instrument measuring high school students' perceived values of learning chemistry. Based on Stuckey et al.'s model and SEVT, students' perceptions were classified into six groups: Intrinsic value, attainment value, utility value, social value, epistemic value, and aesthetic value. The confirmatory factor analysis showed that the six-factor solution produced an adequate model fit. Consistent with previous findings, different types of values were positively and significantly correlated with each other (Fong, Kremer, Cox, & Lawson, 2021; Gaspard et al., 2018; Luttrell et al., 2010; Perez et al., 2019; Trautwein et al., 2012). What is more, students' perceived values of learning chemistry were positively and significantly correlated with their chemistry scores (Gong et al., 2023).

From a theoretical perspective, this study gives a comprehensive conceptualization of the value construct through integrating definitions of values in SEVT with those in Stuckey et al.'s model. Whereas most psychological research on the subjective value has focused on intrinsic value, attainment value, and utility value, findings of this study indicate that epistemic value, aesthetic value, and social value also matter in science education. Descriptive statistics have also showed that Chinese high school students perceived higher levels of social value and epistemic value of learning and similar levels of intrinsic value, attainment value, utility value, and aesthetic value. A wind range of epistemic values are expressed in the pursuit of achieving science curriculum goals including rational thinking, accuracy, creativity, and empiricism (Allchin, 1998; Corrigan et al., 2015; Park, Wu, & Erduran, 2020). For example, the epistemic aspect of learning science is clearly stated in the Framework for K-12 Science Education: "Epistemic knowledge is knowledge of the constructs and values that are intrinsic to science. Students need to understand what is meant, for example, by an observation, a hypothesis, an inference, a model, a theory, or a claim, and be able to distinguish among them" (NRC, 2012, p. 79). Similarly, core literacies listed in Chinese high school chemistry curriculum standards are closely associated with the epistemic value, including making connections between the macroscopic level and microscopic level, notions of change and balance, reasoning from the evidence, modeling, scientific inquiry, and creativity. Considering the significant role of getting high grades in school systems, high school teachers pay the most attention to the epistemic aspect of learning chemistry. Besides epistemic value, core literacies in Chinese high school chemistry curriculum standards also highlight the importance of cultivating students' social responsibilities, which is related to the subconstruct of social value. Compared with other individual-level values, social value describes the value of learning chemistry at the macroscopic level – the improvement of country, society, environments, and human beings. The highest level of social value suggests that students can distinguish between personal-level values and societal-level values, which is consistent with previous findings that students may perceive some chemistry topics (e.g., water) as more relevant at the macroscopic level but less relevant at the personal level (Salonen, Kärkkäinen, & Keinonen, 2018).

Aesthetic value, a newly added component of the value construct, refers to personal appreciation of beauty derived from macroscopic natural phenomena (e.g., flame reactions) and microscopic structures (e.g., symmetry) or the potential of scientific ideas or ways of thinking for helping humans understand or explore the truth or the world (Girod, Rau, & Schepige, 2003; Müller, 2003). The appreciation of the beauty can promote students' interests in learning chemistry (i.e., intrinsic value) (Wickman, Prain, & Tytler, 2022) while the appreciation of the power or potential of the science subject is closely associated with students' perceived epistemic value of learning chemistry, which provides explanations for its high correlations with intrinsic value and epistemic value.

Regarding the other three types of values delineated in SEVT (i.e., intrinsic value, attainment value, and utility value), findings of this study are consistent with previous research, which supports the validity of applying SEVT in different disciplinary and cultural contexts. What is more, qualitative open responses provide a more holistic approach to understand students' perceived values in such contexts. Specifically, the utility value of learning chemistry is further divided into academic utility, practical utility, occupational utility, relational utility, and research utility. Except for the research utility which describes the value of learning chemistry for achieving research purposes (e.g., creating new materials), these subcategories roughly align with Gaspard et al.'s (2017, 2020) distinctions. The utility value is so closely associated with attainment value that sometimes researchers combine them as the importance value (Jacobs, Lanza, Osgood, Eccles, & Wigfield, 2002; Watt Shapka, Morris, Durik, Keating, & Eccles, 2012). Survey items measuring attainment value usually focus on the general importance of performing well on tasks or achieving goals listed in utility value. In this study, students' open responses demonstrate that students' attainment value is expressed in the satisfaction of learning needs, the sense of accomplishment, and the demonstration of individuals' intelligence and life's value. All the preceding information provides rich sources for education researchers to develop more appropriate and accurate items for measuring the attainment value.

From a practical perspective, high school teachers may think about what are effective pedagogical strategies for fostering students' intrinsic, attainment, utility, social, epistemic, and aesthetic value of learning chemistry. Regarding the utility value, researchers usually develop school interventions in two different ways: The first one is to directly provide students with utility-related information of specific techniques (Shechter, Durik, Miyamoto, & Harackiewicz, 2011) while the other is to encourage students to make arguments for the utility of one subject and reflect on typical arguments (Gaspard et al., 2015; Hulleman & Harackiewicz, 2009; Hulleman, Godes, Hendricks, & Harackiewicz, 2010). Regarding the aesthetic value, Chesky and Wolfmeyer (2015) have called for enhancing STEM disciplines' potential for fostering aesthetic value. Ling, Xiang, Chen, Zhang, and Ren (2020) suggested using photography activities in four stages to help students perceive, appreciate, explore, and create the beauty of chemistry. Regarding the intrinsic value and social value, chemistry teachers can create open-inquiry activities or incorporate social scientific issues in daily instruction or enhance the connection with chemists, engineers in chemistry industries, or university faculties through extracurricular activities (De Jong & Talanquer, 2015). Only after students perceive personal relevance or values of learning chemistry are there opportunities for them to retain continuous engagement and choose chemistry-related majors or jobs in the future.

Limitations and Future Directions

This study provides a holistic approach to understanding and measuring high school students' perceived values of learning chemistry. However, there are several limitations that should be taken into consideration. First, whereas I used a large sample to investigate students' perceptions of values of learning chemistry, the sample was limited to two private high schools in northern China. Future research could adopt various methods (e.g., interviews) and collect data from more diverse populations (e.g., rural and urban schools, middle school students) to investigate students' thoughts in depth and reexamine the survey's construct validity. Second, since our recent work has examined how six different types of values correlated with students' chemistry scores (Gong et al., 2023), future research could adopt structural equation modeling to examine how these different types of values predict students' academic behaviors, motivational or affective variables, and future career choices, which provide more robust evidence for the survey's predictive validity. In addition, our recent work has indicated there existed gender differences in Chinese high school students' perceived values of learning chemistry (Gong et al., 2023). Therefore, it is worthwhile further investigating potential cultural differences and why some values are more salient for specific student populations than others.

Appendix

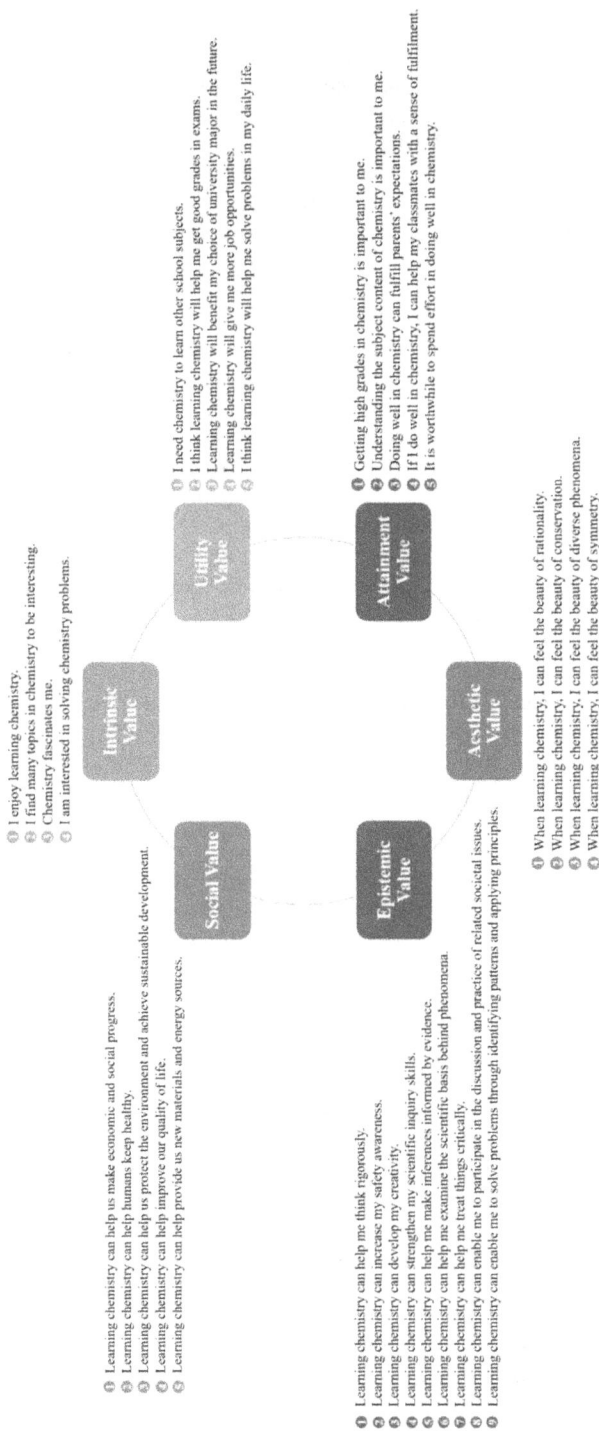

Intrinsic Value
- ① I enjoy learning chemistry.
- ② I find many topics in chemistry to be interesting.
- ③ Chemistry fascinates me.
- ④ I am interested in solving chemistry problems.

Utility Value
- ① I need chemistry to learn other school subjects.
- ② I think learning chemistry will help me get good grades in exams.
- ③ Learning chemistry will benefit my choice of university major in the future.
- ④ Learning chemistry will give me more job opportunities.
- ⑤ I think learning chemistry will help me solve problems in my daily life.

Attainment Value
- ① Getting high grades in chemistry is important to me.
- ② Understanding the subject content of chemistry is important to me.
- ③ Doing well in chemistry can fulfill parents' expectations.
- ④ If I do well in chemistry, I can help my classmates with a sense of fulfilment.
- ⑤ It is worthwhile to spend effort in doing well in chemistry.

Aesthetic Value
- ① When learning chemistry, I can feel the beauty of rationality.
- ② When learning chemistry, I can feel the beauty of conservation.
- ③ When learning chemistry, I can feel the beauty of diverse phenomena.
- ④ When learning chemistry, I can feel the beauty of symmetry.

Social Value
- ① Learning chemistry can help us make economic and social progress.
- ② Learning chemistry can help humans keep healthy.
- ③ Learning chemistry can help us protect the environment and achieve sustainable development.
- ④ Learning chemistry can help improve our quality of life.
- ⑤ Learning chemistry can help provide us new materials and energy sources.

Epistemic Value
- ① Learning chemistry can help me think rigorously.
- ② Learning chemistry can increase my safety awareness.
- ③ Learning chemistry can develop my creativity.
- ④ Learning chemistry can strengthen my scientific inquiry skills.
- ⑤ Learning chemistry can help me make inferences informed by evidence.
- ⑥ Learning chemistry can help me examine the scientific basis behind phenomena.
- ⑦ Learning chemistry can help me treat things critically.
- ⑧ Learning chemistry can enable me to participate in the discussion and practice of related societal issues.
- ⑨ Learning chemistry can enable me to solve problems through identifying patterns and applying principles.

Figure A6 Survey items for measuring six types of values.

References

Allchin, D. (1998). Values in science and science education. In B. Fraser & K. Tobin (Eds.), *International handbook of science education* (pp. 1083–1092). Dordrecht: Kluwer.

Braun, V., & Clarke, V. (2006). Using thematic analysis in psychology. *Qualitative Research in Psychology*, *3*(2), 77–101.

Chesky, N. Z., & Wolfmeyer, M. R. (2015). *Philosophy of STEM education: A critical investigation*. New York: Palgrave Macmillan.

Corrigan, D., Cooper, R., & Keast, S. (2015). The role of values in chemistry education. In Eilks, I., Hofstein, A. (Eds.), *Relevant chemistry education: From theory to practice* (pp. 101–118). Netherlands: Sense Publisher.

De Jong, O., & Talanquer, V. (2015). Why is it relevant to learn the big ideas in chemistry at school? In I. Eilks & A. Hofstein (Eds.), *Relevant chemistry education – From theory to practice* (pp. 11–31). Rotterdam: Sense Publishers.

Dreamson, N. (2018). Culturally inclusive global citizenship education: Metaphysical and nonwestern approaches. *Multicultural Education Review*, *10*(2), 75–93.

Eccles, J. S. (2005). Subjective task value and the Eccles et al. model of achievement-related choices. In A. J. Elliot, & C. S. Dweck (Eds.), *Handbook of competence and motivation* (pp. 105–121). London: Guildford Press.

Eccles, J. S., Adler, T. F., Futterman, R., Goff, S. B., Kaczala, C. M., Meece, J. L., & Midgley, C. (1983). Expectancies, values and academic behaviors. In J. T. Spence (Ed.), *Achievement and achievement motivation* (pp. 75–146). San Francisco: W. H. Freeman.

Eilks, I., & Hofstein, A. (2015). From some historical reflections on the issue of relevance of chemistry education towards a model and an advance organizer–a prologue. In I. Eilks, & A. Hofstein (Eds.), *Relevant chemistry education: From theory to practice* (pp. 1–10). Rotterdam: Sense Publishers.

Fong, C. J., Kremer, K. P., Hill-Troglin Cox, C., & Lawson, C. A. (2021). Expectancy-value profiles in math and science: A person-centered approach to cross-domain motivation with academic and STEM-related outcomes. *Contemporary Educational Psychology*, *65*(March), 101962.

Gaspard, H., Dicke, A. L., Flunger, B., Brisson, B. M., Häfner, I., Nagengast, B., & Trautwein, U. (2015). Fostering adolescents' value beliefs for mathematics with a relevance intervention in the classroom. *Developmental Psychology*, *51*(9), 1226–1240.

Gaspard, H., Häfner, I., Parrisius, C., Trautwein, U., & Nagengast, B. (2017). Assessing task values in five subjects during secondary school: Measurement structure and mean level differences across grade level, gender, and academic subject. *Contemporary Educational Psychology*, *48*, 67–84.

Gaspard, H., Jiang, Y., Piesch, H., Nagengast, B., Jia, N., Lee, J., & Bong, M. (2020). Assessing students' values and costs in three countries: Gender and age differences within countries and structural differences across countries. *Learning and Individual Differences*, *79*, 101836.

Gaspard, H., Wigfield, A., Jiang, Y., Nagengast, B., Trautwein, U., & Marsh, H. W. (2018). Dimensional comparisons: How academic track students' achievements are related to their expectancy and value beliefs across multiple domains. *Contemporary Educational Psychology*, *52*, 1–14.

Girod, M., Rau, C., & Schepige, A. (2003). Appreciating the beauty of science ideas: Teaching for aesthetic understanding. *Science Education*, *87*(4), 574–587.

Gong, X., Bergey, B. W., Jin, Y., Mao, K., & Cheng, Y. (2023). Gender differences in high school students' perceived values and costs of learning chemistry. *Chemistry Education Research and Practice*, *24*(1), 203–216.

Hu, L., & Bentler, P. M. (1999). Cutoff criteria for fit indexes in covariance structure analysis: Conventional criteria versus new alternatives. *Structural Equation Modeling*, *6*, 1–55.

Hulleman, C. S., Godes, O., Hendricks, B. L., & Harackiewicz, J. M. (2010). Enhancing interest and performance with a utility value intervention. *Journal of Educational Psychology*, *102*(4), 880–895.

Hulleman, C. S. & Harackiewicz, J. M. (2009). Promoting interest and performance in high school science classes. *Science*, *326*(5958), 1410–1412.

Jacobs, J. E., Lanza, S., Osgood, D. W., Eccles, J. S., & Wigfield, A. (2002). Changes in children's self-competence and values: Gender and domain differences across grades one through twelve. *Child Development*, *73*(2), 509–527.

Ling, Y., Xiang, J., Chen, K., Zhang, J., & Ren, H. (2020). Integrating aesthetics education into chemistry education: Students perceive, appreciate, explore, and create the beauty of chemistry in Scientific Photography Activity. *Journal of Chemical Education*, *97*(6), 1556–1565.

Luo, X. (2021). *Students' perceived values of learning chemistry: Using the organic-compound unit as the example* (unpublished master dissertation). Tianjin Normal University, Tianjin, China.

Luttrell, V. R., Callen, B. W., Allen, C. S., Wood, M. D., Deeds, D. G., & Richard, D. C. S. (2010). The mathematics value inventory for general education students: Development and initial validation. *Educational and Psychological Measurement*, *70*(1), 142–160.

Müller, A. (2003). Chemistry: The beauty of symmetry. *Science*, *300*(5620), 749–750.

National Research Council (2012). *A Framework for K-12 Science Education: Practices, Crosscutting Concepts, and Core Ideas*. Washington, DC: The National Academy Press.

Nieswandt, M. (2007). Student affect and conceptual understanding in learning chemistry. *Journal of Research in Science Teaching*, *44*(7), 908–937. https://doi.org/10.1002/tea.20169

Osborne, J., & Collins, S. (2001). Pupils' views of the role and value of the science curriculum: A focus-group study. *International Journal of Science Education*, *23*(5), 441–467.

Park, W., Wu, J. Y., & Erduran, S. (2020). Investigating the epistemic nature of STEM: Analysis of science curriculum documents from the USA using the family resemblance approach. In Anderson, J., & Li, Y. (Eds.), *Integrated approaches to STEM education: An international perspective* (pp. 137–155). Dordrecht: Springer.

Perez, T., Dai, T., Kaplan, A., Cromley, J. G., Brooks, W. D., White, A. C., Mara, K. R., & Balsai, M. J. (2019). Interrelations among expectancies, task values, and perceived costs in undergraduate biology achievement. *Learning and Individual Differences*, *72*(April), 26–38.

Salonen, A., Kärkkäinen, S., & Keinonen, T. (2018). Career-related instruction promoting students' career awareness and interest towards science learning. *Chemistry Education Research and Practice*, *19*(2), 474–483.

Shechter, O. G., Durik, A. M., Miyamoto, Y., & Harackiewicz, J. M. (2011). The role of utility value in achievement behavior: the importance of culture. *Personality and Social Psychology Bulletin*, *37*(3), 303–317.

Shi, X. (2021). *Students' perceived values of learning chemistry: Using the unit of iron and metallic materials as the example* (unpublished master dissertation). Tianjin Normal University, Tianjin, China.

Stuckey, M., Hofstein, A., Mamlok-Naaman, R., & Eilks, I. (2013). The meaning of 'relevance' in science education and its implications for the science curriculum. *Studies in Science Education, 49*(1), 1–34.

The Ministry of Education of the People's Republic of China. (2020). *Chemistry Curriculum Standards of Compulsory Education*. Beijing: Beijing Normal University Press.

Trautwein, U., Marsh, H. W., Nagengast, B., Lüdtke, O., Nagy, G., & Jonkmann, K. (2012). Probing for the multiplicative term in modern expectancy-value theory: A latent interaction modeling study. *Journal of Educational Psychology, 104*(3), 763–777.

Wang, M. T., & Degol, J. (2013). Motivational pathways to STEM career choices: Using expectancy-value perspective to understand individual and gender differences in STEM fields. *Developmental Review*, 33, 304–340.

Watt, H. M. G., Shapka, J. D., Morris, Z. A., Durik, A. M., Keating, D. P., & Eccles, J. S. (2012). Gendered motivational processes affecting high school mathematics participation, educational aspirations, and career plans: A comparison of samples from Australia, Canada, and the United States. *Developmental Psychology, 48*(6), 1594–1611.

Wickman, P. O., Prain, V., & Tytler, R. (2022). Aesthetics, affect, and making meaning in science education: An introduction. *International Journal of Science Education, 44*(5), 717–734.

Wigfield, A., & Eccles, J. S. (2020). 35 years of research on students' subjective task values and motivation: A look back and a look forward. In *Advances in motivation science* (Vol. 7, pp. 161–198). Elsevier.

Wigfield, A., Tonks, S., & Klauda S. T. (2009). Expectancy-value theory. In K. R. Wentzel & A. Wigfield (Eds.), *Handbook of motivation at school* (pp. 55–75). New York, NY: Routledge.

7 The Relationship between Students' Motivational Beliefs of Chemistry Learning and Future Course Choices

Introduction

As described in Chapter 6, high school students' perceived values of learning chemistry are represented in multiple aspects. At the macroscopic level, the chemistry subject plays an important role in promoting the economic and sustainable development of society. At the personal level, as its content knowledge is closely associated with modern sciences of material, life, environment, information, and energy (The Ministry of Education, 2020), chemistry education in high school directly influences students' future course or career choices in either a positive or a negative way. However, it has been reported that students around the world are unwilling to choose majors or careers related to chemistry (Avargil, Kohen, & Dori, 2020; Wang, Rocabado, Lewis, & Lewis, 2021). A similar pattern also emerges in the Chinese context, especially after the national reform of the college entrance examination in which chemistry has become an optional course in high school (Wang & Yang, 2021). Investigating factors that influence students' course and career choices of chemistry can help researchers and educators better develop school interventions to increase student motivations to learn chemistry. In this study, I employed the situated expectancy-value theory (SEVT) as the theoretical framework because it systemically describes the mechanism of how motivational factors (e.g., self-efficacy beliefs, perceived values, and costs of tasks) associate with students' academic achievement and choice behaviors (Wigfield & Eccles, 2020). At present, numerous studies have used it to examine students' achievement, choices, or retention in STEM–related fields in Western contexts (Chow, Eccles, & Salmela-Aro, 2012; Fong, Kremer, Cox, & Lawson, 2021; Guo, Marsh, Parker, Morin, & Dicke, 2017; Guo, Wang, Ketonen, Eccles, & Salmela-Aro, 2018; Nagy, Trautwein, Baumert, Köller, & Garrett, 2006; Watt et al., 2012). However, only a few studies have investigated how students' motivational beliefs, including self-efficacy and perceived values, relate to academic choice behaviors in Eastern countries (e.g., Chang, 2015; Wan, 2021). What is more, most studies with SEVT have assessed positive aspects of task values (e.g., intrinsic value, attainment value, and utility value) and ignored the negative aspect (e.g., costs).

DOI: 10.4324/9781032649665-11

Recent research has called for including cost as a separate construct in order to explain students' academic outcomes or behaviors more accurately (Barron & Hulleman, 2015; Jiang, Kim, & Bong, 2020; Perez et al., 2019). To address these research gaps, the main purpose of this study was to investigate the factor structures of chemistry self-efficacy beliefs, perceived values, and costs of learning chemistry among Chinese high school students and identify their relations with course choices in the national college entrance examination. The results of this study can provide evidence to support the cross-cultural application of SEVT and shed light on how educators can strengthen students' willingness to select the chemistry subject and related careers through designing interventions targeting motivational beliefs.

Literature Review

The Conceptualization of Motivational Beliefs in SEVT

The SEVT model postulates that individuals' expectations of success and subjective task value are two important predictors of achievement-related choices and performance (Wigfield & Eccles, 2020). Among them, the expectation of success refers to the individual's beliefs in his or her abilities to accomplish academic goals or achieve desired task outcomes. Previous research has used different terms (e.g., self-concept, self-efficacy) to measure students' ability beliefs. As self-efficacy is more domain-specific than other variables (Bandura, 2006), I selected it for evaluating students' expectancy beliefs in this study. Similarly, the subjective value is also task specific and refers to perceived positive or negative aspects of achievement tasks (Eccles et al., 1983). Accordingly, the construct of subjective task value consists of three positive components and one negative component: Intrinsic value, utility value, attainment value, and relative cost. As previous chapters have elaborated definitions and functions of three positive components (i.e., intrinsic value, utility value, and attainment value), I focus on the negative construct – relative cost – and its joint contribution with values in predicting academic outcomes in this chapter.

In contrast to positive aspects of subjective task value, the construct of cost and its relations with academic outcomes has received less attention until recently, specifically in STEM fields (Barron & Hulleman, 2015; González & Paoloni, 2015). It is defined as what is lost or given up or suffered when engaging in particular activities (Wigfield & Eccles, 2020). Similar to subjective task value, perceived cost is also multi-dimensional and researchers have conceptualized four different types of costs: ***Effort cost, opportunity cost, ego cost*** and ***emotional cost.*** Correspondingly, ***effort cost*** refers to individuals' perceptions of excessive effort taken to complete the task (e.g., does my chemistry homework require more effort than I want to spend on it?); ***opportunity cost*** refers to the loss of other opportunities due to engaging in one activity (e.g., do I choose to do my chemistry homework or play with friends?); ***ego cost*** refers to ego threats associated with potential failures (e.g., does the poor performance in chemistry make me feel incompetent?); *emotional cost* refers to negative emotions emerged from engaging in the task (e.g., do taking chemistry exams make me feel stressed and anxious?) (Eccles et al., 1983; Flake, Barron, Hulleman, McCoach, & Welsh, 2015; Jiang et al., 2020; Luttrell et al., 2010; Perez et al., 2019). The preceding definitions provide the basis for understanding and exploring theoretical structures of perceived costs.

However, the relation between positive aspects of subjective task value and relative cost has been debated in the extant literature (Barron & Hulleman, 2015). Some researchers thought that the construct of cost was in parallel with the other three positive components and therefore calculated a mean composite task value by combining the preceding four components (e.g., Buehl & Alexander, 2005). Other researchers suggested treating cost as a distinguishable variable from values (Barron & Hulleman, 2015; Jiang et al., 2020) because it has explained additional variance in academic achievement and behaviors beyond expectancy and value (Jiang, Rosenzweig, & Gaspard, 2018). This disagreement indicates that more empirical work should be done to investigate the relations between positive aspects of subjective task value and relative cost.

Roles of Motivational Beliefs in Predicting Academic Outcomes

Consistent with theoretical hypotheses in SEVT model, many empirical studies with different student populations have confirmed that students' motivations beliefs including self-efficacy beliefs, perceived values, and costs are closely associated with student engagement, achievement, and academic-related choices in different cultural contexts (e.g., Berndt & Miller, 1990; Bong, 2001; Galla, Amemiya, & Wang, 2018; Gaspard et al., 2018; Jiang et al., 2020; Luo, Ng, Lee, & Aye, 2016; Trautwein et al., 2012; Guo et al., 2017; Guo, Marsh, Parker, Morin, & Yeung, 2015; Guo, Wang, Ketonen, Eccles, & Salmela-Aro, 2018). Among them, most studies have included variables measuring expectation of success and all or some components of task values while only a few studies have taken the role of cost into consideration. Regarding expectation of success, results have showed that students' science self-efficacy positively predicted their science competencies (i.e., identifying scientific issues, explaining phenomena scientifically, and using scientific evidence) and behavioral tendency to learn science in classrooms (e.g., Chang, 2015; Wan, 2021). Regarding perceived values, three types of positive subjective task values (i.e., intrinsic value, attainment value, and utility value) have been reported to be positively associated with students' science course grades or task performance in Western and Eastern contexts (e.g., Chang, 2015; Gaspard et al., 2018; González & Paoloni, 2015; Perez et al., 2019). In contrast, perceived costs typically have negative associations with students' academic choice, persistence, and achievement after controlling for positive task values. For example, Jiang et al. (2018) reported that among South Korean middle school students' perceived costs including effort, opportunity, ego, and emotional cost significantly and negatively predicted math achievement after controlling for task value. Perez,

Cromley, and Kaplan (2014) found that effort cost and opportunity cost positively predicted students' intentions of leaving STEM majors. They further examined how undergraduate students' perceived costs mediated the relation between expectancy beliefs and biology achievement. Results showed that the relation between expectancy beliefs and biology achievement was strongest when effort cost was low while the relation was lowest when effort cost was high (Perez et al., 2019). Compared to self-efficacy beliefs and positive subjective value, which have been extensively examined in previous research, the role of cost in predicting students' academic outcomes has received researchers' attention until recently. What is more, complex relations among expectancy beliefs, subjective task value, and cost indicate that more research is needed, particularly in different disciplinary and cultural contexts.

The Present Study

Although many studies have examined theoretical hypotheses in SEVT in different domains, there is a lack of empirical research addressing students' perceived value or cost of learning chemistry, particularly in Eastern cultural contexts (Gaspard et al., 2020). Therefore, the purpose of the current study was to investigate factor structures of perceived values and costs of learning chemistry among Chinese high school students. As cultural contexts shape the meaning of values and the cost construct is under-researched in the literature (Jiang et al., 2020), it is worthwhile to investigate meaning and assess different aspects of values and costs in non-Western settings (Wigfield, Tonks, & Klauda, 2009; Gaspard et al., 2020). Further, I examined how chemistry self-efficacy beliefs, perceived values and costs of learning chemistry predicted students' intentions of selecting chemistry in the college entrance examination. Such empirical evidence contributes to better understanding how different types of costs and values related to students' academic-related choices in different domains. Based on the basis of SEVT, I proposed the following hypotheses for the two research questions. First, students' perceived values and costs of learning chemistry were multidimensional; second, self-efficacy beliefs and perceived values were positive predictors of students' course choices while perceived costs were negative predictors of students' course choices.

Method

Participants and Procedure

The data was collected in the 2018–2019 academic year from one public school located in a northern province in China. Before the data collection, informed consent was obtained from school principals and classroom teachers. At the beginning of the survey, students were also informed that participation was completely voluntary and anonymous. The collected data was treated confidentially and only be used for research purposes. Participants were 477 high school students with an average age of 15.7 years old (SD = .90). Demographically, 53.9% were 10th grade and 45.9% were 11th grade (the remaining 0.2% did not indicate grade), and 3.4% were from minority groups. The distribution of participants by gender was 49.9% male and 49.5% female (the remaining 0.6% did not indicate gender). It took students about 10 minutes to complete the paper- and-pencil survey.

Instruments

Survey items were written in Chinese and measured high school students' feelings toward the subject of chemistry. The translation and back-translation procedure was adopted to guarantee the equivalence of meaning of survey items that were originally developed in English (Brislin, 1970). The survey included two parts: Demographic information (e.g., grade level, gender, age, and race/ethnicity) and Likert scale items. In the second part, a response scale ranging from 1 (strongly disagree) to 6 (strongly agree) was used to measure students' perceptions of chemistry self-efficacy beliefs, values, and costs of learning chemistry. At the end of the survey, one item measured students' future intentions of choosing chemistry as the subject tested in the college entrance examination. All measures demonstrated good to excellent internal-consistency reliability (see the following reliability coefficients).

- *Chemistry Self-efficacy.* Five items were adapted from the self-efficacy subscale of the Chinese version of the Motivated Strategies for Learning Questionnaire (MLSQ) developed by Rao and Sachs (1999) to measure students' chemistry self-efficacy beliefs. The reliability coefficient was α = .90. One example item was: "I expect to do well in chemistry classes".
- *Perceived Values of Learning Chemistry.* Ten items were used to measure students' perceived values of learning chemistry in three categories: Intrinsic value, attainment value, and utility value. Specifically, two items measuring intrinsic value were adapted from the Mathematics Value Inventory (MVI) developed by Luttrell et al. (2010). Survey items were modified for the chemistry subject by changing "math" to "chemistry". One example item was: "I enjoy learning chemistry". Three items measuring utility value were adapted from the mathematics value scale used in Trends in International Mathematics and Science Study (TIMSS) (Olson, Martin, & Mullis, 2008). One example item was: "I think learning chemistry will help me in my daily life". Among five items measuring attainment value, two items were adapted from the subscale of need for high achievement in MVI (Luttrell et al., 2010) and three items were adapted from TIMSS2019. One example item was: "Understanding the ideas taught in chemistry classes is important to me". The reliability coefficient was α = .917.
- *Perceived Costs of Learning Chemistry.* Eight items were adopted from the cost scale developed by Jiang, Kim, and Bong (2020) to measure students' perceived costs of learning chemistry including emotional cost, ego cost,

and effort cost. The reliability coefficient of the cost instrument was $\alpha = .726$. One example item measuring emotional cost was: "Studying chemistry makes me feel stressed".

Overview of Data Analyses

The Mplus was used to conduct the analysis of survey data. First, I conducted a series of confirmatory factor analyses to examine the factor structure of Likert items measuring three motivational variables (i.e., chemistry self-efficacy beliefs, perceived values and costs of learning chemistry). Maximum likelihood was used to estimate the model fitting. The goodness of fit was assessed with the following indices: The chi-square test of model fit (χ^2), the degree of freedom (df), the root mean square error of approximation (RMSEA), the comparative fit index (CFI), the Tucker-Lewis index (TLI), and the standardized root mean square residuals (SRMR) (Hu &Bentler, 1999). Based on existing criteria (Byrne, 2010; Hu & Bentler, 1999; Kline, 2010), values of CFI and TLI above .90 indicate a good model fit to the data while values of RMSEA and SRMR below .08 indicate a reasonable model fit. Second, a structural equation model (SEM) was applied to examine how different factors extracted from three motivational variables predicted students' future intentions of selecting chemistry as the subject tested in the college entrance examination.

Results

The Measurement Model

I first used Mplus to conduct a series of confirmatory factor analyses with maximum likelihood estimation to evaluate the validity and reliability of latent constructs. The indices for different models are listed in Table 7.1. For the

Table 7.1 Model fit statistics for confirmatory factor analyses

Model	χ^2	df	CFI	TLI	RMSEA	SRMR
Self-efficacy						
One factor	**20.6**	**5**	**.988**	**.977**	**.081**	**.016**
Value						
One factor	509.125	35	.838	.792	.169	.064
Two factors	**256.168**	**34**	**.924**	**.900**	**.118**	**.051**
Three factors	235.082	32	.931	.903	.116	.049
Cost						
One factor	749.349	20	.482	.275	.279	.169
Three factors	**59.253**	**17**	**.970**	**.951**	**.073**	**.040**
All						
Six Factors	**768.162**	**215**	**.917**	**.902**	**.074**	**.066**
2nd order	1075.488	222	.872	.854	.090	.157

chemistry self-efficacy construct, a one-factor structure showed a good model fit (χ^2 = 20.6, df = 5, CFI = .988, TLI = .977, RMSEA = .081, SRMR = .016). For constructs of perceived values and costs of learning chemistry, I first tested a one-factor model that included all value- and cost-related items. The indices showed that the one-factor model did not fit the data well (χ^2 = 1970.602, df = 135, CFI = .616, TLI = .565, RMSEA = .170, SRMR = .126). Therefore, I decided to treat values and costs as separate constructs. For the value construct, I tested one-factor model, two-factor model (i.e., intrinsic value and importance value), and three-factor model (i.e., intrinsic value, attainment value, and utility value) in sequence. The best solution was determined by comparing a set of fit statistics between different models and considerations that indicated the adequacy of the model relative to more parsimonious models. Despite the fact that the value of RMSEA was greater than 0.10, the fit of a two-factor model was acceptable because the other three fit indices were in the expected range. Therefore, it was concluded that the two-factor model of the value construct fitted the data reasonably (χ^2 = 256.168, df = 34, CFI = .924, TLI = .900, RMSEA = .118, SRMR = .051). For the cost construct, the three-factor solution including effort cost, emotional cost, and ego cost (χ^2 = 59.253, df = 17, CFI = .970, TLI = .951, RMSEA = .073, SRMR = .040) showed a much better model fit than the one-factor solution (χ^2 = 749.349, df = 20, CFI = .482, TLI = .275, RMSEA = .279, SRMR = .169).

Before testing hypothesized relations between variables, I also compared model fitting indices between a comprehensive model and a second-order model. In the comprehensive model, there were six latent variables (i.e., self-efficacy, intrinsic value, importance value, effort cost, emotional cost, and ego cost). In the second-order model, first-order factors representing intrinsic value and importance value were set to load on a second-order factor representing the value construct while first-order factors representing effort cost, emotional cost, and ego cost were set to load on the other second-order factor representing the cost construct. Table 7.2 listed factor loadings of the first-order factors

Table 7.2 Second-order confirmatory factor analysis: factor loadings of the first-order factors and intercorrelations between second-order factors

Factor	Self-efficacy	Value	Cost
Intrinsic value		.911**	
Importance value		.868**	
Effort cost			.445**
Emotional cost			.231**
Ego cost			.973**
Self-efficacy	–		
Value	.820**	–	
Cost	.086	.378**	–

** p-value significant at the 0.01 level.

Table 7.3 Descriptive statistics of six latent factors

Factor	Self-efficacy	Intrinsic value	Importance value	Effort cost	Emotional cost	Ego cost
Mean	3.36	3.78	3.92	3.14	2.74	3.20
SD	0.88	1.14	0.84	1.02	1.03	1.18
Skewness	−0.28	−0.86	−1.13	−0.09	0.10	−0.29
Kurtosis	−0.04	−0.02	1.85	−0.43	−0.63	−0.70

(i.e., intrinsic value, importance value, effort cost, emotional cost, and ego cost) on the two second-order factors and intercorrelations between second-order factors (i.e., self-efficacy, value, and cost). Factor loadings ranged from .231 (between emotional cost and cost) to .973 (between ego cost and cost). Results showed that the second-order model (χ^2 = 1075.488, df = 222, CFI = .872, TLI = .854, RMSEA = .090, SRMR = .157) had a weaker fit than the six-factor model (χ^2 = 768.162, df = 215, CFI = .917, TLI = .902, RMSEA = .074, SRMR = .066) (see Table 7.1). Therefore, I adopted the six-factor model to examine the influence of self-efficacy beliefs, values, and costs on students' course choices in the college entrance examination. Descriptive statistics of six latent factors are listed in Table 7.3.

Structural Equation Model

The structural equation model was conducted to examine the initial hypotheses regarding the relationships among self-efficacy beliefs, intrinsic value, importance value, effort cost, emotional cost, ego cost, and course choices. The final model showed a good fit to the data (χ^2 = 849.492, df = 232, CFI = .913, TLI = .896, RMSEA = .075, SRMR = .066). The standardized factor loadings, representing the relationships between the indicators and latent variables, were all statistically significant (p < .001) and ranged from .55 to .94, indicating the adequate validity of latent variables in the measurement model.

Standardized paths for the final model are showed in Figure 7.1. As showed, four out of six direct paths were statistically significant. Self-efficacy beliefs (β = .09, p = .214) and effort cost (β = −.02, p = .698) did not significantly predict students' chemistry course choices. Intrinsic value (β = .27, p = 0.002) and ego cost (β = .12, p = 0.019) positively and significantly predicted students' intentions of selecting chemistry as the subject tested in the college entrance examination while emotional cost (β = −.13, p = .041) negatively and significantly predicted their course choices. In other words, high school students who perceived higher intrinsic value, importance value, and ego cost were more likely to select chemistry as the subject tested in the college entrance examination while students with high emotional cost were reluctant to select the chemistry subject. The aforementioned latent variables explained 54.5% of variance in the outcome variable. Covariances between six latent factors are listed in Table 7.4. The majority of covariances were significant except for the covariance between

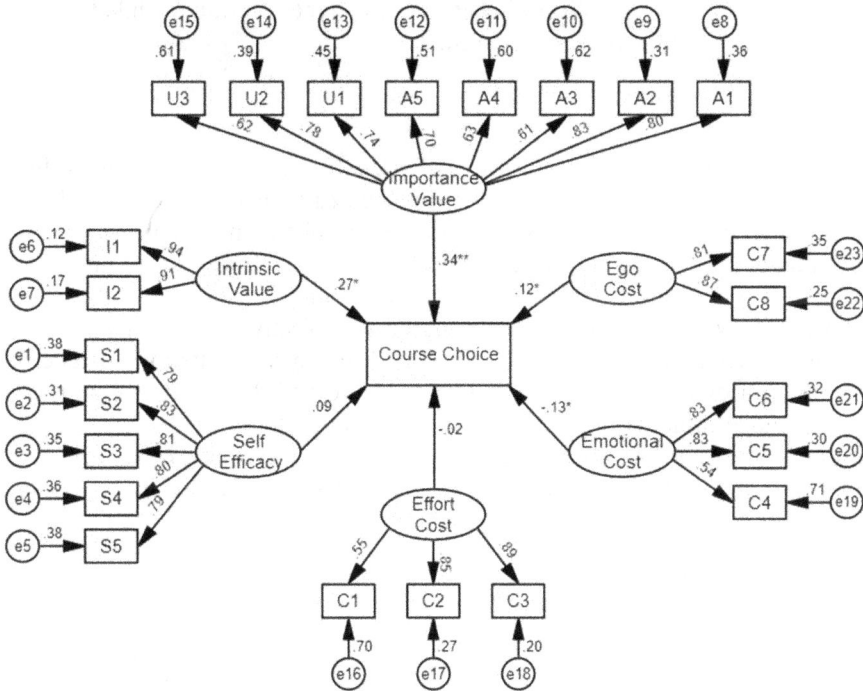

Figure 7.1 Structural equation model.

Table 7.4 Covariances among six latent variables

Factor	1	2	3	4	5	6
1. Self-efficacy	–					
2. Intrinsic value	.78**	–				
3. Importance value	.67**	.79**	–			
4. Effort cost	.26**	.28**	.45**	–		
5. Emotional cost	−.61**	−.55**	−.30**	.13*	–	
6. Ego cost	.08	.26**	.43**	.42**	.28**	–

Note: Factor loadings and coefficients are standardized; For reasons of clarity, covariances among six latent variables are not included in the model.
* *p* < .05.
** *p* < .001.

self-efficacy and ego cost (*p* = .131). As expected, self-efficacy and two types of values were positively and strongly related to each other, ranging from .67 to .79. Three types of costs were also positively and significantly related to each other, ranging from .13 to .42. In terms of covariances among self-efficacy beliefs, perceived values, and costs, results showed that effort cost and ego cost were positively related to self-efficacy beliefs and two types of values while the emotional cost showed the opposite pattern.

Discussion

This study identified factor structures of self-efficacy beliefs, perceived values, and costs in the chemistry domain and examined their associations with students' tendency to select chemistry as the tested subject in the college entrance examination. Findings suggested that chemistry self-efficacy was unidimensional while perceived values and costs of learning chemistry were multidimensional. What is more, positive aspects of subjective task value including intrinsic value and importance value were positively associated with students' course choices while different types of costs exerted different effects on students' course choices. This study adds to the extant literature by including self-efficacy beliefs, task values, and perceived costs together to examine their unique predictive patterns with the outcome variable in the chemistry domain.

Factor Structures of Self-efficacy Beliefs, Perceived Values and Costs

Compared to self-efficacy beliefs, students' perceived values and costs of learning chemistry were multidimensional. Regarding the value construct, findings of this study indicated a two-factor structure. Specifically, intrinsic value was a relatively separate construct while utility value and attainment value were related and emerged as the importance value. It was consistent with findings of another study which showed that items measuring utility value and attainment value grouped together and substantially overlapped in non-metric multidimensional scaling maps (Gong, Bergey, Jin, Mao, & Cheng, 2023). It also provided empirical evidence for the rationality of combining utility value and attainment value and terming the summary construct as importance value (Durik, Vida, & Eccles, 2006; Jacobs, Lanza, Osgood, Eccles, & Wigfield, 2002; Watt et al., 2012). Regarding the cost construct, though SEVT has theoretically hypothesized it as a negative component of subjective task value, empirical results of confirmatory factor analyses showed that the cost was a separate and distinct variable from values, which were consistent with previous findings (Gaspard et al., 2015; Luttrell et al., 2010).

Existing research has operationalized constructs of value and cost in different ways. Some studies focused on one specific positive component of value (e.g., Chouinard & Roy, 2008; Dietrich, Dicke, Kracke, & Noack, 2015; Guo et al., 2017; Luo et al., 2016; Nagy et al., 2006; Wang et al., 2021) while other studies treated value as a single construct by combining different components and calculating mean composite score (Buehl & Alexander, 2005; Jacobs et al., 2002; Perez et al., 2014; Safavian & Conley, 2016). Therefore, instead of combining values and costs together as a composite construct, cost should be treated as a construct independent of values (Wigfield, Rosenzweig, & Eccles, 2017; Jiang et al., 2020). What is more, this study confirms the multidimensionality of cost and provides empirical support for the validity of the three-factor structure including emotional cost, ego cost, and effort cost (Wigfield et al., 2017). Results of this study reaffirm the necessity of using confirmatory factor analyses to identify distinct components of the value construct and the cost construct before examining their power in explaining academic outcomes (Conley, 2012; Gaspard et al., 2015; Trautwein et al., 2012).

The Influence of Three Motivational Variables on Chemistry Course Choices

Finds from the current study confirm unique roles of values and costs in predicting students' academic-related choices and persistence (Eccles, 2009; Guo et al., 2017; Perez et al., 2014). However, the role of self-efficacy beliefs is not significant, which is not consistent with previous research. Some researchers found that competence-related beliefs, or perceived task values or costs positively predicted students' coursework aspirations and achievement simultaneously (Guo et al., 2017; Jiang et al., 2018) while other researchers reported that value beliefs lost predictive power when expectancy beliefs were taken into consideration (Trautwein et al., 2012). There are several possible explanations for inconsistent findings regarding relations between three motivational variables and academic outcomes. On the one hand, researchers often include different independent variables or operationalize these variables in different ways, which directly influence the statistical significance and amount of variance explained in outcome variables. On the other hand, researchers may employ different terms to measure academic outcomes including achievement, engagement, persistence, participation, aspirations, and course intentions (e.g., Watt et al., 2012). Previous research suggests that specific motivational variables are more closely associated with some variables than others. For example, it has been theoretically hypothesized that expectancy beliefs are more related to achievement (Wigfield, Tonks, & Klauda, 2009) while value beliefs are important predictors of academic effort (Nagengast et al., 2011; Trautwein, Lüdtke, Schnyder, & Niggli, 2006). In addition, perceived costs are more related to academic choices than to achievement (Eccles et al., 1983). These findings emphasize the need to further examine their complex relations and provide additional evidence that the cost construct predicts students' choices or intentions above and beyond values (Perez et al., 2014). Based on these findings, it can be concluded that cost should be considered as a separate and distinct construct from values (Jiang et al., 2020; Wigfield, Rosenzweig, & Eccles, 2017).

What is more, components of perceived values and costs of learning chemistry predict students' course choices in different ways. Specifically, importance value was the strongest predictor, followed by intrinsic value while the two components of costs predicted students' intentions in opposite directions: Emotional cost was a significant and negative predictor while ego cost was a significant and positive predictor. Some findings align with previous research on the role of perceived values in influencing achievement and intentions but also contradict previous findings when it comes to the relations of different types of costs with outcome variables. For example, the negative path from emotional cost to course choices is compatible with previous findings that emotional cost is closely associated with students' avoidance intentions (e.g., dropping out of course or school) (de la Varre, Irvin, Jordan, Hannum, & Farmer, 2014; Jiang et al., 2020; Zhu & Chen, 2013). Its negative role is also supported by the evidence that emotional cost was negatively related to self-efficacy beliefs, intrinsic value, and importance value while positively related to effort cost and ego cost. By contrast, ego cost, which refers to the self-worth threat resulting from potential task failures (Eccles et al., 1983), positively predicted students' intentions of selecting the chemistry subject in the college entrance examination. Relations between ego cost and outcome variables may depend on cultural contexts and independent variables under investigation. For example, Jiang et al. (2020) found that ego cost may demonstrate interpretative relevance but the path from ego cost to academic outcomes was not significant after controlling for general cost. Markus and Kitayama (1991) have described the salience of ego threats among Eastern students due to the significant role of academic performance in shaping the sense of identity. Therefore, students who perceive higher ego cost of learning chemistry might approach this challenge more positively and be more likely to select the chemistry subject. In fact, students' course choices are the result of the compromise between internal motivational factors (e.g., self-efficacy beliefs, values, and costs) and external policy factors. For example, policies related to college entrance examination stipulate that students who plan to pursue majors in material science, environmental science, and medical science have to select chemistry as the tested subject. Consequently, even though students may perceive effort cost and ego cost during the process of learning chemistry, students may choose to overcome potential obstacles and try their best to achieve their career and life goals. Just as an old Chinese proverb says, "Diligence redeems stupidity and practice makes perfect". This also explains why effort cost was not a significant predictor of high school students' course choices. In addition, effort cost was found to be more related to academic achievement than opportunity and psychological cost in the domain of science (Perez et al., 2019). In summary, findings of this study contribute to understanding the structure of costs and provide empirical evidence for the theoretical hypothesis that specific dimensions of costs (i.e., effort cost, emotional cost, and ego cost) play different roles in predicting students' various academic outcomes after accounting for their self-efficacy beliefs and perceived values (Jiang et al., 2018; Perez et al., 2014), which are still underexplored in existing literature (Jiang et al., 2020; Perez et al., 2019).

Conclusions, Implications, and Limitation

In this study, I first adopted confirmatory factor analyses to examine factor structures of self-efficacy beliefs, perceived values, and costs of learning chemistry. Results showed that the value was a second-factor construct (i.e., intrinsic value and importance value) while the cost was a three-factor construct (i.e., effort cost, emotional cost, and ego cost). I further adopted structural equation model to examine their associations with students' course choices. Results showed that two types of values and ego cost were significant and positive predictors while emotional cost was a significant and negative predictor. As existing studies that examine the role of cost in influencing students' academic performance often treat it as a general construct, findings of this study confirm the necessity and rationality of differentiating between its subdimensions. What is more, this study supports the theoretical hypothesis that an expectancy-value-cost approach can explain more variance in students' academic outcomes. However, it is intriguing to find that students' chemistry self-efficacy beliefs lost its predictive power when other variables were included. In other words, students who perceived higher values of learning chemistry may still select chemistry as the subject tested in the college entrance examination regardless of whether they have much confidence in their abilities to learn chemistry. It is also worth noting that effects of emotional cost, ego cost, and effort cost on course choices may depend on specific cultural values or norms in different contexts. Considering complex relations between self-efficacy beliefs and subcomponents of values and costs, I recommend that future researchers pay more attention to roles of different components of costs in predicting academic achievement and behaviors and their interactions with competence beliefs and values. Based on these findings, I suggest that school interventions targeting to improve students' interests in learning chemistry should not only improve students' perceived relevance of chemistry through reflections and writing prompts but also deeply evaluate how individuals perceive costs of learning chemistry and their attitudes toward them. As cultural values may shape students' perceptions of values and costs, more cross-cultural investigations and interventions are needed to identify effective strategies that can promote students' decisions of selecting the chemistry subject and future retentions in chemistry-related majors and jobs.

Besides these aforementioned implications, several limitations to this study should be noted. First, the survey data were collected at a single measurement point which could not demonstrate causal relations among self-efficacy beliefs, values, and costs. Future research could conduct a longitudinal study to examine directions and interactions among these variables. Second, the self-report data were collected from one high school in northern China, which decreased the generalizability of findings. Future research could collect data from more participants covering different regions and different school types. Similarly, this study only focused on the domain of chemistry. As students' course choices in the college entrance examination are shaped by various internal and external factors, its generalizability to other domains should be further examined. Finally, as this study only used one item to measure students' course choices, future research could develop multiple items to measure this variable more accurately.

References

Avargil, S., Kohen, Z., & Dori, Y. J. (2020). Trends and perceptions of choosing chemistry as a major and a career. *Chemistry Education Research and Practice, 21*(2), 668–684. https://doi.org/10.1039/c9rp00158a

Bandura, A. (2006). Guide for constructing self-efficacy scales. In F. Pajares, & T. Urdan (Eds.), *Self-efficacy beliefs of adolescents* (pp. 307–337). Greenwich, CT: Information Age Publishing.

Barron, K. E., & Hulleman, C. S. (2015). Expectancy-value-cost model of motivation. In J. S. Eccles, & K. Salmelo-Aro (Eds.), *Motivational psychology* (2nd ed.), pp. 261–271). Amsterdam, Netherlands: Elsevier.

Berndt, T. J., & Miller, K. E. (1990). Expectancies, Values, and Achievement in Junior High School. *Journal of Educational Psychology, 82*(2), 319–326. https://doi.org/10.1037/0022-0663.82.2.319

Bong, M. (2001). Between- and within-domain relations of academic motivation among middle and high school students: Self-efficacy, task-value, and achievement goals. *Journal of Educational Psychology, 93*(1), 23–34. https://doi.org/10.1037/0022-0663.93.1.23

Brislin, R. W. (1970). Back-translation for cross-cultural research. *Journal of Cross-Cultural Psychology, 1*, 185–216.

Buehl, M. M., & Alexander, P. A. (2005). Motivation and performance differences in students' domain-specific epistemological belief profiles. *American Educational Research Journal, 42*, 697–726. https://doi.org/10.3102/00028312042004697.

Byrne, B. (2010). *Structural equation modeling with AMOS. Basic concepts, applications, and programming* (2nd ed.). New York: Routledge.

Chang, Y. (2015). Science motivation across Asian Countries: Links among future-oriented motivation, self-efficacy, task values, and achievement outcomes. *Asia-Pacific Education Researcher, 24*(1), 247–258. https://doi.org/10.1007/s40299-014-0176-3

Chouinard, R., & Roy, N. (2008). Changes in high-school students' competence beliefs, utility value and achievement goals in mathematics. *British Journal of Educational Psychology, 78*(1), 31–50. https://doi.org/10.1348/000709907X197993

Chow, A., Eccles, J. S., & Salmela-Aro, K. (2012). Task value profiles across subjects and aspirations to physical and IT-related sciences in the United States and Finland. *Developmental Psychology, 48*(6), 1612–1628.

Conley, A. M. M. (2012). Patterns of motivation beliefs: Combining achievement goal and expectancy-value perspectives. *Journal of Educational Psychology, 104*(1), 32–47. https://doi.org/10.1037/a0026042

de la Varre, C., Irvin, M. J., Jordan, A. W., Hannum, W. H., & Farmer, T. W. (2014). Reasons for student dropout in an online course in a rural K-12 setting. *Distance Education, 35*(3), 324–344. https://doi.org/10.1080/01587919.2015.955259.

Dietrich, J., Dicke, A. L., Kracke, B., & Noack, P. (2015). Teacher support and its influence on students' intrinsic value and effort: Dimensional comparison effects across subjects. *Learning and Instruction, 39*, 45–54. https://doi.org/10.1016/j.learninstruc.2015.05.007

Durik, A. M., Vida, M., & Eccles, J. S. (2006). Task values and ability beliefs as predictors of high school literacy choices: A developmental analysis. *Journal of Educational Psychology, 98*(2), 382–393. https://doi.org/10.1037/0022-0663.98.2.382

Eccles, J. S. (2009). Who am I and what am I going to do with my life? Personal and collective identities as motivators of action. *Educational Psychologist, 44*, 78–89. http://dx.doi.org/10.1080/00461520902832368.

Eccles, J. S., Adler, T. F., Futterman, R., Goff, S. B., Kaczala, C. M., Meece, J. L., & Midgley, C. (1983). Expectancies, values, and academic behaviors. In J. T. Spence (Ed.). *Achievement and achievement motives: Psychological and sociological approaches* (pp. 75–146). San Francisco, CA: W. H. Freeman and Company.

Flake, J. K., Barron, K. E., Hulleman, C., McCoach, B. D., & Welsh, M. E. (2015). Measuring cost: The forgotten component of expectancy-value theory. *Contemporary Educational Psychology, 41*, 232–244. https://doi.org/10.1016/j.cedpsych.2015.03.002

Fong, C. J., Kremer, K. P., Cox, C. H. T., & Lawson, C. A. (2021). Expectancy-value profiles in math and science: A person-centered approach to cross-domain motivation with academic and STEM-related outcomes. *Contemporary Educational Psychology, 65*, 101962. https://doi.org/10.1016/j.cedpsych.2021.101962

Galla, B. M., Amemiya, J., & Wang, M. T. (2018). Using expectancy-value theory to understand academic self-control. *Learning and Instruction, 58*(April), 22–33. https://doi.org/10.1016/j.learninstruc.2018.04.004

Gaspard, H., Dicke, A. L., Flunger, B., Schreier, B., Häfner, I., Trautwein, U., & Nagengast, B. (2015). More value through greater differentiation: Gender differences in value beliefs about math. *Journal of Educational Psychology, 107*(3), 663–677. https://doi.org/10.1037/edu0000003

Gaspard, H., Jiang, Y., Piesch, H., Nagengast, B., Jia, N., Lee, J., & Bong, M. (2020). Assessing students' values and costs in three countries: Gender and age differences within countries and structural differences across countries. *Learning and Individual Differences, 79*, 101836. https://doi.org/10.1016/j.lindif.2020.101836

Gaspard, H., Wigfield, A., Jiang, Y., Nagengast, B., Trautwein, U., & Marsh, H. W. (2018). Dimensional comparisons: How academic track students' achievements are related to their expectancy and value beliefs across multiple domains. *Contemporary Educational Psychology, 52*, 1–14. https://doi.org/10.1016/j.cedpsych.2017.10.003

Gong, X., Bergey, B. W., Jin, Y., Mao, K., & Cheng, Y. (2023). Gender differences in high school students' perceived values and costs of learning chemistry. *Chemistry Education Research and Practice, 24*(1), 203–216. https://doi.org/10.1039/d2rp00169a

González, A., & Paoloni, P. V. (2015). Perceived autonomy-support, expectancy, value, metacognitive strategies and performance in chemistry: A structural equation model in undergraduates. *Chemistry Education Research and Practice, 16*(3), 640–653. https://doi.org/10.1039/c5rp00058k

Guo, J., Marsh, H. W., Parker, P. D., Morin, A. J. S., & Dicke, T. (2017). Extending expectancy-value theory predictions of achievement and aspirations in science: Dimensional comparison processes and expectancy-by-value interactions. *Learning and Instruction, 49*, 81–91. https://doi.org/10.1016/j.learninstruc.2016.12.007

Guo, J., Marsh, H. W., Parker, P. D., Morin, A. J. S., & Yeung, A. S. (2015). Expectancy-value in mathematics, gender and socioeconomic background as predictors of achievement and aspirations: A multi-cohort study. *Learning and Individual Differences, 37*, 161–168. https://doi.org/10.1016/j.lindif.2015.01.008

Guo, J., Wang, M. T., Ketonen, E. E., Eccles, J. S., & Salmela-Aro, K. (2018). Joint trajectories of task value in multiple subject domains: From both variable- and pattern-centered perspectives. *Contemporary Educational Psychology, 55*, 139–154. https://doi.org/10.1016/j.cedpsych.2018.10.004

Hu, L., & Bentler, P. M. (1999). Cutoff criteria for fit indexes in covariance structure analysis: Conventional criteria versus new alternatives. *Structural Equation Modeling, 6*, 1–55.

Jacobs, J. E., Lanza, S., Osgood, D. W., Eccles, J. S., & Wigfield, A. (2002). Changes in children's self-competence and values: Gender and domain differences across grades one through twelve. *Child Development, 73*, 509–527. https://doi.org/10.1111/1467-8624.00421

Jiang, Y., Kim, S., & Bong, M. (2020). The role of cost in adolescent students' maladaptive academic outcomes. *Journal of School Psychology, 83*, 1–24. https://doi.org/10.1016/j.jsp.2020.08.004

Jiang, Y., Rosenzweig, E. Q., & Gaspard, H. (2018). An expectancy-value-cost approach in predicting adolescent students' academic motivation and achievement. *Contemporary Educational Psychology, 54*(June), 139–152. https://doi.org/10.1016/j.cedpsych.2018.06.005

Kline, R. B. (2010). *Principles and practice of structural equation modeling* (3rd ed.). New York, NY: Guilford.

Luttrell, V. R., Callen, B. W., Allen, C. S., Wood, M. D., Deeds, D. G., & Richard, D. C. S. (2010). The Mathematics Value Inventory for General Education Students: Development and Initial Validation. *Educational and Psychological Measurement, 70*(1), 142–160. https://doi.org/10.1177/0013164409344526

Luo, W., Ng, P. T., Lee, K., & Aye, K. M. (2016). Self-efficacy, value, and achievement emotions as mediators between parenting practice and homework behavior: A control-value theory perspective. *Learning and Individual Differences, 50*, 275–282. https://doi.org/10.1016/j.lindif.2016.07.017

Markus, H. R., & Kitayama, S. (1991). Culture and the self: Implications for cognition, emotion, and motivation. *Psychological Review, 98*, 224–253. https://doi.org/10.1037/0033-295X.98.2.224

Nagengast, B., Marsh, H. W., Scalas, L. F., Xu, M., Hau, K. T., & Trautwein, U. (2011). Who took the "X" out of expectancy–value theory? A psychological mystery, a substantive-methodological synergy, and a cross-national generalization. *Psychological Science, 22*, 1058–1066. https://doi.org/10.1177/0956797611415540

Nagy, G., Trautwein, U., Baumert, J., Köller, O., & Garrett, J. (2006). Gender and course selection in upper secondary education: Effects of academic self-concept and intrinsic value. *Educational Research and Evaluation, 12*(4), 323–345. https://doi.org/10.1080/13803610600765687

Olson, J. F., Martin, M. O., & Mullis, I. V. S. (2008). TIMSS 2007 technical report. Chestnut Hill, MA: TIMSS & PIRLS International Study Center.

Perez, T., Dai, T., Kaplan, A., Cromley, J. G., Brooks, W. D., White, A. C., Mara, K. R., & Balsai, M. J. (2019). Interrelations among expectancies, task values, and perceived costs in undergraduate biology achievement. *Learning and Individual Differences, 72*(April), 26–38. https://doi.org/10.1016/j.lindif.2019.04.001

Perez, T., Cromley, J. G., & Kaplan, A. (2014). The role of identity development, values, and costs in college STEM retention. *Journal of Educational Psychology, 106*(1), 315–329. https://doi.org/10.1037/a0034027

Rao, N., & Sachs, J. (1999). Confirmatory factor analysis of the Chinese version of the motivated strategies for learning questionnaire. *Educational and Psychological Measurement, 59*(6), 1016–1029.

Safavian, N., & Conley, A. M. (2016). Expectancy-value beliefs of early-adolescent Hispanic and non-Hispanic youth: Predictors of mathematics achievement and enrollment. *AERA Open*, *2*, 1–17.

The Ministry of Education of the People's Republic of China. (2020). *Revised High School Chemistry curriculum standards*. Beijing: People's Education Press. (In Chinese)

Trautwein, U., Lüdtke, O., Schnyder, I., & Niggli, A. (2006). Predicting homework effort: Support for a domain-specific, multilevel homework model. *Journal of Educational Psychology*, *98*, 438–456. https://doi.org/10.1037/0022-0663.98.2.438

Trautwein, U., Marsh, H. W., Nagengast, B., Lüdtke, O., Nagy, G., & Jonkmann, K. (2012). Probing for the multiplicative term in modern expectancy-value theory: A latent interaction modeling study. *Journal of Educational Psychology*, *104*(3), 763–777. https://doi.org/10.1037/a0027470

Wan, Z. H. (2021). Exploring the effects of intrinsic motive, utilitarian motive, and self-efficacy on students' science learning in the classroom using the Expectancy-Value Theory. *Research in Science Education*, *51*(3), 647–659. https://doi.org/10.1007/s11165-018-9811-y

Wang, Y., Rocabado, G. A., Lewis, J. E., & Lewis, S. E. (2021). Prompts to promote success: Evaluating utility value and growth mindset interventions on general chemistry students' attitude and academic performance. *Journal of Chemical Education*, *98*(5), 1476–1488. https://doi.org/10.1021/acs.jchemed.0c01497

Watt, H. M. G., Shapka, J. D., Morris, Z. A., Durik, A. M., Keating, D. P., & Eccles, J. S. (2012). Gendered motivational processes affecting high school mathematics participation, educational aspirations, and career plans: A comparison of samples from Australia, Canada, and the United States. *Developmental Psychology*, *48*(6), 1594–1611. https://doi.org/10.1037/a0027838

Wigfield, A., & Eccles, J. S. (2020). 35 years of research on students' subjective task values and motivation: A look back and a look forward. In A. Elliot (Vol. Ed.), *Advances in Motivation Science* (1st ed., Vol. 7). New York: Elsevier Inc. https://doi.org/10.1016/bs.adms.2019.05.002

Wigfield, A., Rosenzweig, E. Q., & Eccles, J. (2017). Achievement values: Interactions, interventions, and future directions. In A. Elliot, C. Dweck, & D. Yeager (Eds.), *Handbook of competence and motivation: Theory and application* (2nd ed.). New York, NY: Guilford Press.

Wigfield, A., Tonks, S., & Klauda, S. L. (2009). Expectancy–value theory. In K. R. Wentzel, & A. Wigfield (Eds.), *Handbook of motivation at school* (pp. 55–75). New York, NY: Routledge.

Wang, Q., & Yang, Y. (2021). A study on the factors influencing chemistry subject selection under the background of new college entrance examination: Based on the survey from Chongqing's implementation of "3+1+2" program. *Journal of Teacher Education*, *8*(5), 70–77. (In Chinese)

Zhu, X., & Chen, A. (2013). Motivational cost aspects of physical education in middle school students. *Educational Psychology*, *33*, 465–481. https://doi.org/10.1080/01443410.2013.785043.

Index

Pages in *italics* refer to figures and pages in **bold** refer to tables.

For Product Safety Concerns and Information please contact our EU
representative GPSR@taylorandfrancis.com
Taylor & Francis Verlag GmbH, Kaufingerstraße 24, 80331 München, Germany

9 781032 649702